Decision Making in the
U.S. Department of Energy's Environmental Management
Office of Science and Technology

Committee on Prioritization and Decision Making in the
Department of Energy Office of Science and Technology

Board on Radioactive Waste Management
Commission on Geosciences, Environment, and Resources
National Research Council

NATIONAL ACADEMY PRESS
Washington, D.C.

NOTICE: The project that is the subject of this report was approved by the Governing Board of the National Research Council, whose members are drawn from the councils of the National Academy of Sciences, the National Academy of Engineering, and the Institute of Medicine. The members of the committee responsible for the report were chosen for their special competences and with regard for appropriate balance.

Support for this study was provided by the U.S. Department of Energy under Grant No. DE-FC01-94EW54069. All opinions, findings, conclusions, or recommendations expressed herein are those of the authors and do not necessarily reflect the views of the U.S. Department of Energy.

On the cover: Photographs of Office of Science and Technology technologies used at Department of Energy Environmental Management sites. Clockwise from upper-right: containers of mixed waste stored prior to their remediation at the East Tennessee Technology Park at the Oak Ridge Reservation; the VecLoader™ used for deactivation and decommissioning at the Fernald Site's Plant 1; double-wall high-level waste tanks encased in concrete underground at the Hanford Site; Room 152 at the Rocky Flats Environmental Technology Site, showing glove boxes, conduit, and piping used for nuclear materials; and the Radioactive Waste Management Complex at the Idaho National Engineering and Environmental Laboratory.

International Standard Book Number 0-309-06347-7

Additional copies of this report are available from:

National Academy Press
2101 Constitution Avenue, N.W.
Box 285
Washington, DC 20055
800-624-6242
202-334-3313 (in the Washington Metropolitan Area)
http://www.nap.edu

COMMITTEE ON PRIORITIZATION AND DECISION MAKING IN THE DEPARTMENT OF ENERGY OFFICE OF SCIENCE AND TECHNOLOGY

RAYMOND G. WYMER, *Chair*, Oak Ridge National Laboratory, Oak Ridge, Tennessee (retired)
ALLEN G. CROFF, Oak Ridge National Laboratory, Oak Ridge, Tennessee
MARY R. ENGLISH, University of Tennessee, Knoxville, Tennessee
THOMAS M. JOHNSON, LFR Levine-Fricke, Emeryville, California
DUNDAR F. KOCAOGLU, Portland State University, Portland, Oregon
MICHAEL MENKE, Value Creation Associates, Redwood City, California
GEORGE L. NEMHAUSER, Georgia Institute of Technology, Atlanta, Georgia
LINDA WENNERBERG, Environmental Business Strategies, Cambridge, Massachusetts
EDWIN L. ZEBROSKI, Elgis Consulting, Inc., Sunnyvale, California

Consultants

THOMAS A. COTTON, JK Research Associates, Inc., Arlington, Virginia
ROBERT GIORDANO, Giordano and Associates, Saratoga Springs, New York
DETLOF VON WINTERFELDT, University of Southern California, Los Angeles, California

NRC Staff

THOMAS E. KIESS, Study Director
ROBIN L. ALLEN, Senior Project Assistant
MATTHEW BAXTER-PARROTT, Project Assistant

The National Academy of Sciences is a private, nonprofit, self-perpetuating society of distinguished scholars engaged in scientific and engineering research, dedicated to the furtherance of science and technology and to their use for the general welfare. Upon the authority of the charter granted to it by the Congress in 1863, the Academy has a mandate that requires it to advise the federal government on scientific and technical matters. Dr. Bruce Alberts is president of the National Academy of Sciences.

The National Academy of Engineering was established in 1964, under the charter of the National Academy of Sciences, as a parallel organization of outstanding engineers. It is autonomous in its administration and in the selection of its members, sharing with the National Academy of Sciences the responsibility for advising the federal government. The National Academy of Engineering also sponsors engineering programs aimed at meeting national needs, encourages education and research, and recognizes the superior achievements of engineers. Dr. William A. Wulf is president of the National Academy of Engineering.

The Institute of Medicine was established in 1970 by the National Academy of Sciences to secure the services of eminent members of appropriate professions in the examination of policy matters pertaining to the health of the public. The Institute acts under the responsibility given to the National Academy of Sciences by its congressional charter to be an adviser to the federal government and, upon its own initiative, to identify issues of medical care, research, and education. Dr. Kenneth Shine is president of the Institute of Medicine.

The National Research Council was organized by the National Academy of Sciences in 1916 to associate the broad community of science and technology with the Academy's purposes of furthering knowledge and advising the federal government. Functioning in accordance with general policies determined by the Academy, the Council has become the principal operating agency of both the National Academy of Sciences and the National Academy of Engineering in providing services to the government, the public, and the scientific and engineering communities. The Council is administered jointly by both Academies and the Institute of Medicine. Dr. Bruce Alberts and Dr. William A. Wulf are the chairman and vice-chairman, respectively, of the National Research Council.

Acknowledgments

In carrying out this study the committee was briefed during many meetings in Washington, D.C. and at the U.S. Department of Energy (DOE) sites where the Office of Science and Technology (OST) Focus Areas were managed. A large number of DOE program managers and site management and operations personnel prepared and made presentations to the committee and answered the many questions posed by committee members. The committee is grateful for the patience and courtesy shown it in the course of its inquiries. The committee is especially grateful to Clyde Frank, Gerald Boyd, Jef Walker, and Carolyn Davis for the time they gave to the committee out of their very busy schedules. Discussions with them were candid and productive with regard to providing the committee with insights into how the decision-making process in OST has evolved and matured over the approximately nine years of OST's existence.

In addition to the discussions with DOE, committee members held discussions with representatives from industry to learn how they deal with the problems of deciding what technologies to develop to stay competitive and be profitable. Although the goals and environment in which the companies operate are quite different from those of OST, nonetheless there are lessons to be learned from them that are applicable to OST. The committee is grateful to representatives of the Gas Research Institute, particularly Ronald Edelstein; the Electric Power Research Institute, particularly Steven Gehl; and DuPont, particularly H. Dale Martin, for their time and advice.

The committee is also appreciative of the interest and participation of D. Warner North, who served as a committee liaison to the BRWM.

Finally, it would not have been possible for the committee to function without the capable assistance of the National Research Council staff. Robin Allen and Matthew Baxter-Parrott provided invaluable meeting and committee support and carried out the many behind-the-scenes tasks without which the committee could not have functioned. The committee is especially grateful to Tom Kiess whose unflagging enthusiasm and hard work in many areas, such as contacting DOE and site personnel and others, prodding committee members to complete their writing tasks, and contributing to the writing, was essential to completion of this study.

Acknowledgement of Reviewers

This report has been reviewed in draft form by individuals chosen for their diverse perspectives and technical expertise, in accordance with procedures approved by the National Research Council (NRC) Report Review Committee. The purpose of this independent review is to provide candid and critical comments that will assist the institution in making the published report as sound as possible and to ensure that the report meets institutional standards for objectivity, evidence, and responsiveness to the study charge. The content of the review comments and draft manuscript remain confidential to protect the integrity of the deliberative process. We wish to thank the following individuals for their participation in the review of this report:

Paul Barton, United States Geological Survey (retired)
Robert Budnitz, Future Resources Associates, Inc.
Paul Busch, Malcolm Pirnie, Inc.
Radford Byerly, Jr., University Corporation for Atmospheric Research (retired)
Walter Haerer, Montgomery Watson
Kai N. Lee, Williams College
Jane Long, University of Nevada at Reno.
Thomas Saaty, University of Pittsburgh (retired)
Marcia Williams, Putnam, Hayes, and Bartlett, Inc.

While the individuals listed above have provided constructive comments and suggestions, it must be emphasized that responsibility for the final content of this report rests entirely with the authoring committee and the institution.

Preface

In 1996, the National Research Council (NRC) of the National Academy of Sciences, National Academy of Engineering, and Institute of Medicine was asked by the U.S. Department of Energy (DOE) Office of Science and Technology (OST) to study the processes by which OST makes decisions on which technologies to support in connection with cleaning up DOE defense sites, and to make recommendations on how these processes might be improved. A Committee on Prioritization and Decision Making in the DOE OST was established to carry out this assignment. The task has proven to be especially difficult because OST's decision-making processes are not static, but are changing almost continuously, and in very significant ways. The committee has attempted to deal with this circumstance by stressing that OST's decision-making process is evolutionary and should be viewed in that light. The committee has also stressed that there are general approaches to decision making that can be adopted to advantage, regardless of the details of the process.

The committee entertained no preconceived ideas about the appropriate organizational structure to carry out research and development to meet the needs of DOE-Office of Environmental Management (EM). The committee firmly believes that there is a research and development function to be performed in cleaning up DOE sites and intends that the recommendations in this report be applicable to this function irrespective of the organizational structure or the program office charged with carrying out the work.

The work of this committee is closely related to the work of other NRC committees, especially the Committee on the DOE OST's Peer Review Program that is addressing peer review within OST. To avoid duplication of effort, peer review of project work (a major focus of that committee study) is not dealt with in detail in this report, which instead examines other steps in the decision-making process suitable for reviews.

Information gathering for this report was terminated in April 1998.

Raymond G. Wymer, Chair
Oak Ridge, Tennessee

Contents

Executive Summary

The U.S. Department of Energy (DOE) Environmental Management (EM) program is responsible for radioactive waste management and environmental remediation at DOE sites.[1] Within the DOE-EM program, the Office of Science and Technology (OST) has a mission to develop technology to facilitate these activities by reducing their technical risks and/or costs. OST funds research, development, and demonstration (RD&D) projects, and those that result in viable technologies are available for use by site contractors who are under the direct management of the other DOE-EM offices.

SCOPE AND CONTENT OF THIS REPORT

This National Research Council (NRC) report examines the prioritization and decision-making processes of OST. The major recommendations developed by the committee conducting this study are summarized and organized below around the four decision process issues raised in the committee's statement of task (see Box ES.1). Since the committee found that OST's decision processes are inextricably linked with DOE-EM organizational structure, institutional procedures, and program management, these related program elements are also discussed in this report as necessary.

The broader institutional environment within which OST operates imposes constraints on the OST decision process. Two features of this environment are the other DOE-EM program offices and the federal budget process, both of which impact the decisions of OST program managers. Other significant influences on OST have included changes in top-level DOE-EM priorities and changes in the technical baselines (for example, changing the in-tank precipitation process for cesium removal at the Savannah River plant) that the other EM offices pursue in their site cleanup and waste management activities.

In response to such influences, the programs, prioritization and planning methods, and decision-making processes used by OST have evolved over time. Consequently, the committee sought to evaluate the OST decision-making process in the context of its history and continuing evolution, thereby taking into account changes, improvements, and present directions.

The committee concluded its information-gathering activities in April 1998. Since then, OST has made changes to improve its operations. Nonetheless, this report addresses important characteristic and systemic issues in the RD&D program within DOE-EM, and makes recommendations to further improve OST's operations and the results they achieve.

---

[1] Some of these sites were part of the former nuclear weapons complex, with no future on-site mission except for the remediation of waste inventories, radiologically contaminated structures, and environmental contamination.

BOX ES.1   Statement of Task

This report responds to the task statement reproduced below:

A panel of the CEMT[a] will be appointed to evaluate the effectiveness of the OST decision-making process and make specific recommendations to improve it, if appropriate. In particular, the panel will address the following:

- the appropriateness and effectiveness of decision-making processes currently in use by OST to select, prioritize, and fund RD&D activities, both at sites and at headquarters;
- the technical factors appropriate to consider in the decision-making process for selection, prioritization, and the development of cleanup technologies, and the adequacy with which these factors can be measured;
- recommendations, if appropriate, for improving the decision-making process; and
- the role and importance of effective peer reviews in the decision-making process.

For the latter topic the panel will coordinate its work with the CEMT panel on peer review.[b]

The chair of the NRC appointed a committee of nine scientists and engineers to perform this study (see Appendix K). This committee held information-gathering meetings to obtain input from OST program units and site management personnel. Interviews with individuals and invited speakers at committee meetings provided information from corporations and institutions having technology development problems and activities in some ways similar to those of OST.

---

[a]The Committee on Environmental Management Technologies (CEMT), the parent committee of the decision-making panel, became inactive August 1997, after which time the Board on Radioactive Waste Management provided coordination and oversight of the study panels examining separate facets of the OST program. At that time the decision-making panel was renamed the Committee on Prioritization and Decision Making in the Department of Energy Office of Science and Technology.
[b]This panel, also renamed as a stand-alone committee, was chartered to examine OST's peer review process.

APPROPRIATENESS AND EFFECTIVENESS OF OST DECISION-MAKING PROCESSES

The first issue of the statement of task is "the appropriateness and effectiveness of decision-making processes currently in use by OST to select, prioritize, and fund R&D activities, both at sites and at headquarters." To fulfill its technology development mission in a way that is responsive to DOE-EM site technology needs, OST's major decision process elements can be summarized as (1) identification of high-priority DOE-EM site technology needs, (2) program planning, (3) assessments of technology development proposals and projects, and (4) headquarters oversight. The committee believes that these process elements are appropriate but not fully effective as currently implemented. The recommendations that follow are offered to improve the effectiveness of each of these process elements.

Process Element 1: Identification of High-Priority DOE-EM Site Technology Needs

OST has formed Site Technology Coordination Groups (STCGs) at each major DOE-EM site to interact with local contractor personnel and others to obtain that site's technology needs. The STCGs form statements of these needs that they evaluate and prioritize (i.e., rank or rate) according to a set of STCG-generated criteria. The committee found weaknesses in the STCG structuring of criteria and in the STCG evaluative and prioritization methods. The criteria were different at each site and at some sites were not rigorously constructed. Although improvements to these criteria and methods are possible, the committee instead proposes the following twofold approach for effective identification of high-priority needs.

The first recommendation applies to user-requested and relatively near-term needs. *Recommendation*: **OST should use the best available information on DOE-EM site technology needs as a guide for tailoring program goals and RD&D projects. As one way to acquire this information, OST should establish (or increase) its direct contact with site personnel at the problem-solving and decision-making levels.**

The committee believes that longer-term needs should come from OST's consideration of the functional flowsheets[2] for site remediation that the technology user organizations (i.e., other DOE-EM offices) already develop, use in their planning, and subject to reviews. However, these reviews should ideally include consideration of not only the current "baseline" version of the functional flowsheet, but also alternative functional flowsheets that are designed to deal with uncertainties for process steps with significant risk of technical failure or of underperformance in attaining a cleanup goal. Such alternatives would serve a backup role and produce additional technology needs. At present, OST has no direct role in establishing or reviewing these flowsheets, which are activities conducted by other EM organizations and contractors at the site level. *Recommendation:* **In conjunction with the other DOE-EM offices responsible for site cleanups, OST should participate to the extent possible (e.g., by establishing a role for its contractors) in a review of site remediation functional flowsheets. OST's technology development projects should be responsive to technology needs identified from baseline remediation plans and from alternatives.** This involvement of OST should assist in the derivation of technology needs for remediation plans.

Process Element 2: Program Planning

Each major OST program unit (e.g., a Focus Area or a Crosscutting Program; see Table 1.1 for a complete list) conducts planning activities in order to identify (1) the technical needs of greatest priority for that program and (2) a suitable suite of technology development projects to address these needs. These planning activities are important in directing the allocation of resources and are conducted in different ways; for example, no two Focus Areas have the same approach. This planning occurs in a complex, ever-changing, and politicized environment in which OST has more than one "customer" to satisfy. Other EM offices responsible for site waste management and environmental remediation activities are obvious end users of OST-developed technologies, but the U.S. Congress must also be satisfied that a reasonable fraction of OST products are useful and worth their cost. Furthermore, parts of OST expenditures are congressionally mandated (U.S. Congress, 1994; 1995a-b). Consequently, the type and quality of the information provided to the U.S. Congress (and to other interested review groups) are critically important to OST.

Recommendation: **To aid program planning in this institutional environment, a decision methodology should be employed that is structured using quantifiable attributes wherever**

---

[2]The functional flowsheet for site remediation plans is the planned sequence of process steps that comprise the waste treatment process from the initial waste configuration to the final waste end state.

applicable, but that also allows for managerial flexibility. For decisions involving the allocation of significant resources, OST should institute a decision-making structure wherein projects and/or proposals are evaluated against consistently defined criteria such as project cost, probability of technical success, probability of implementation in field applications, potential cost savings, and human health risk reduction. This structure should be applied broadly throughout the organization, with each OST program unit evaluating projects against the same criteria, which should be quantifiable where possible. An important criterion would be the project's relevance to site activities; that is, all projects should have one or more specific objectives related to the cleanup and waste management goals at sites. For an ongoing OST project, its history of past performance in meeting appropriate developmental objectives would be useful information to gauge its likely future success.

The quantification of project benefits and probabilities of success is necessarily uncertain, but the committee believes that a rigorous attempt to use quantifiable estimates and other available data is preferable to the less structured methods that OST has used in the past. The recognition and evaluation of uncertainties is itself a key part of a decision process.

Any such decision support tool has limitations, particularly if it fails to account for all significant factors;[3] therefore, managerial flexibility to "override the numbers" in occasional situations is necessary. There should also be sufficient managerial flexibility to enable responses to late-breaking needs that are important technology development opportunities.

Recommendation: **Important funding decisions and their rationales should be documented and made publicly available.** The criteria and the decision-making process should be recorded and summarized to serve as a basis for learning from experience and for defending decisions. Records should be kept of the reasoning used in reaching decisions, including method(s) used in evaluations, especially when managerial privilege to "override the numbers" is invoked.

This structured methodology and documentation, in which the goals, factors, and criteria influencing the decision are clearly specified, is a useful way to organize information for the OST managers who make decisions; moreover, it creates a clear and defensible record. The committee's judgment is that the discipline imposed by this methodology would improve the effectiveness of the OST decision-making process by defining how projects relate to top-level goals and projected benefits.

Part of the structure to the recommended decision process is that of estimating cost savings from the use of technologies other than current baseline approaches. Although OST has attempted to obtain a uniform and credible cost-estimating methodology by employing the U.S. Army Corps of Engineers, further steps are needed to establish the credibility of this methodology. *Recommendation*: **OST should do "cost avoidance" (or return on investment) calculations on its more expensive technologies in a more credible manner than was done in past efforts and should communicate the results to potential technology users in the most effective way possible. Initial estimates of costs and benefits should be developed at the inception of large RD&D projects. Refinements of the estimates should be a part of the project as it progresses, and followed up by a comparison of the estimates with the actual incurred costs.**

---

[3]The optimal decision is not always that of selecting the candidate with the highest score from a quantitative evaluation against a set of criteria, particularly if the criteria and scoring system do not adequately represent the complexity of the most relevant considerations. In the absence of a more rigorous method to treat these considerations, the committee suggests that managers be allowed to select candidates that are not necessarily those with the highest scores when there are justifiable and documented reasons to do so.

Process Element 3: Assessments of Technology Development Proposals and Projects

Each OST program unit manages new and ongoing projects. Ongoing projects are continued subject to their relevance to current site technology needs and to their successful review in the "stage-and-gate" system that OST programs use to monitor the status of work in progress. New technology development projects are solicited currently based on statements of high-priority site needs. Principal investigators (PIs) at national laboratories, DOE-EM sites, universities, and industries respond to these solicitations with proposals that are evaluated and scored against a set of criteria, typically by independent review teams. Program managers use these technical inputs, programmatic considerations (e.g., relevance to areas of greatest need and to program goals), and available budget to select the PI proposals to be funded as new activities.

The committee believes that the "stage-and-gate" system to monitor ongoing projects has more development stages, separated by decision points (gates), than necessary for effective management of OST projects. Furthermore, the current use of the "stage-and-gate" system seems to merely track projects. It does not seem to assist effectively in decision making either in (1) the initial decision to select a project for funding or in (2) the decision to terminate a project. *Recommendation*: **OST should use the minimum number of stages and gates needed to track a project and should use peer reviews (NRC, 1997b; 1998b) at key decision points (gates), especially in the selection of a new project.** The independent review teams, if appropriately constituted, provide this initial peer review, but not all OST program units use such review groups.

Before a technology need statement is translated into a solicitation for new technology development projects, the decision process should include a "buy-versus-make" procurement decision of whether the need can be met adequately with technologies already developed and available from the private sector. The present approach to technology procurement—wherein several OST organizational units (specifically, the Industry Program, the Large-Scale Demonstration Projects [LSDPs], and the Accelerated Site Technology Deployment [ASTD] Program) perform some aspects of technology selection and procurement from industry—is cumbersome and duplicative, impairing OST's effectiveness. *Recommendation*: **A better-coordinated, less duplicative, and less cumbersome system should be established for integration of technology procurement activities. Since decisions to develop technologies should be made only if warranted following a "make-or-buy" review, the ability to assess available technology is crucial. These assessments should be done through up-to-date surveys of commercially available technologies that are coordinated across OST organizational units.**

Another component of the decision-making process is the methodology used to evaluate PI proposals received in response to an OST solicitation. For selecting new projects from among many proposals, OST program units typically use simple multi-attribute utility analysis scoring systems. In these constructions, criteria (e.g., safety benefit and technical performance superior to baseline methods) are provided to reviewers, who score a proposal on a numerical scale. These scores are combined with weighting factors to arrive at a composite score for the proposal. In the committee's judgment, these scoring systems are adequate, although improvements are possible.

Process Element 4: Headquarters Oversight

In an organization with several layers of management and program responsibilities, decisions on technology development cut across many boundaries and, to be fully successful, must be accepted by other offices. For OST, these other offices are (1) the DOE-EM technology user programs at DOE sites, which are responsible for both the planning that generates technology needs and the deployment of RD&D results, and (2) the hierarchy of institutions outside the DOE-EM program that are involved in the process that establishes the OST budget. In this institutional environment, OST headquarters oversight has important roles in coordinating the alignment of OST's goals with those of other DOE-EM offices (i.e., the community of technology users), managing tensions that may result from a conflict of goals with another program office or institution, and interacting with other government bodies (e.g., the Office of Management and Budget and the U.S. Congress) to establish the annual budgets of each of the individual OST program units. These considerations lead to the necessity for clear and consistent top-level goals to provide adequate direction to OST's efforts.

The decision-making process within other DOE-EM offices is a "bottom-up" one (DOE, 1997j) relying on site-specific resolutions of cleanup priorities. However, a bottom-up approach does not preclude the use of top-level goals, which have been formulated in the past for these EM offices and for OST and continue to be refined. *Recommendation*: **OST managers, in conjunction with other top-level EM managers,[4] should develop strategic goals and plans that define explicitly the technical problems that OST program units will and will not address.** These strategic goals should be communicated throughout OST and be specific enough to guide decisions. Any such strategic goals for OST should be guided by priorities established within other DOE-EM offices. For example, an area of current importance is the privatization of cleanup efforts at many DOE sites. Since privatization may impact DOE-EM's RD&D activities, top DOE-EM management should establish goals for OST's role in privatized cleanups.

APPROPRIATE TECHNICAL FACTORS AND THE ADEQUACY WITH WHICH THEY CAN BE MEASURED

The second issue of the committee's statement of task is "the technical factors appropriate to consider in the decision-making process for selection, prioritization, and the development of cleanup technologies, and the adequacy with which these factors can be measured." The committee identified three major technical factors: (1) the degree of maturity of a technology, (2) the ease of integration of a technology into a total system as described by functional flowsheets, and (3) the impact of a technology in reducing costs, schedule delays, environmental and health risks, and/or the technical risks of system failure or of inadequate cleanup performance.

The *degree of maturity* of a technology is represented by a developing technology's remaining assumptions and uncertainties that require testing, and can be measured in terms of the estimated time required to achieve a sufficiently large-scale (e.g., pilot-scale) demonstration. *Recommendation*: **The aforementioned gate review of an ongoing project that is ready to progress into a different stage of development should include not just a technical peer review, but also an assessment of the maturity and likelihood of future success of that technology project.[5]** More mature projects have a shorter time to completion and therefore to potential deployment than concepts requiring more research and consequently more time. However, an unbalanced RD&D program having only relatively mature projects would fail to deliver the benefits that can come from investments in projects of a less mature nature that

---

[4]Input from other EM offices is recommended, since OST's goals should be derived from user plans and needs.

[5]The likelihood of success is the combined likelihood of achieving the expected technical performance and achieving at least one field application.

address longer-term needs. Therefore, the decision process should lead to a balanced portfolio of projects of varying degrees of maturity.

The *ease of integration* of a technology into a total cleanup system, as represented by a site remediation plan in functional flowsheet form, depends on the flexibility of the flowsheet in permitting a change in the technical approach to one of the individual remediation steps. In proposing a technology change, a new technology must be more than simply an adequate substitute for the baseline technology it is proposed to replace. The new technology must have significantly better technical performance, less technical risk, and/or lower cost. However, for an OST technology to be integrated into a site remediation plan to meet a specific cleanup need, nontechnical factors also must be addressed. One nontechnical issue is that some remediation approaches are constrained by existing regulatory agreements and cannot be changed without renegotiation. The earlier recommendation for OST to participate in a review of the site remediation functional flowsheets will provide OST with information germane to both the technical and the nontechnical issues identified here.

The intended *impact* of a replacement technology is to reduce costs, schedules, and/or risks of various kinds, and/or to enable a difficult or intractable task to proceed. *Recommendation*: **The gate reviews of the stage-and-gate tracking system should also assess estimations of cost, risk, and schedule reductions, for situations in which this information is available. OST should use the resulting information to terminate projects with insufficient potential to achieve significant impact.**

One way to estimate these cost, risk, and schedule impacts is to identify the number of functional flowsheets needing a technology, the number of needs in each flowsheet, and the importance of each flowsheet need. The technology user community is the source of the information necessary to make quantitative assessments in this way. Such assessed impacts could enable OST to make informed decisions about which site needs to address with technology development projects. OST's current criteria for funding these projects include consideration of a technology's potential applicability at more than one DOE-EM site. Although this guideline is useful, the committee believes that it should be balanced against the opportunity for a technology to be primarily beneficial for a single application or single site. The quantification of technology impacts would enable these (and other) opportunities to be readily identified and would lessen the need for OST to rely upon guidelines.

RECOMMENDATIONS FOR IMPROVEMENT

The third issue of the committee's statement of task is "recommendations, if appropriate, for improving the decision-making process." In addition to the recommendations shown previously, OST's decision making can be enhanced by learning from the practices of the private sector.

Application of Practices in Private-Sector RD&D Decision Making

Industries engaging in environmental decision making and the utility consortia of the Gas Research Institute (GRI) and the Electric Power Research Institute (EPRI) apply some of the practices listed below in settings analogous in some ways to OST's. Therefore, the committee thinks that these practices are relevant to DOE's technology development decisions. *Recommendation*: **OST should adopt, where applicable and appropriate in the OST environment and to the extent practicable, basic principles of private-sector formal decision making and follow-up practices.**

The committee recommends that DOE-OST focus on the following major practices:

- Understand, focus on, and monitor changes in customer needs and requirements.
- Agree on clear and measurable goals.
- Use a formal (i.e., common, consistent, structured, and rational) technology development decision-making process and apply it uniformly.
  - Think strategically (i.e., long-term and high impact).
  - Measure and evaluate to guide resource allocation.
  - Communicate across organizational boundaries (i.e., with technology users).
  - Continually improve the research and development (R&D) management process.
  - Hire the best people possible and maintain expertise.

These descriptors of practices are fashioned to capture one or more related ideas taken from already published work and are offered here to describe the attributes of a successful R&D management environment. Since decision processes exist in the context of an organizational structure, some of the practices apply to RD&D management, whereas others apply to the decision process itself. OST management already uses many of these practices to some extent, but the committee recommends that OST re-examine the degree and effectiveness of their use.

Chapter 5 describes possible applications of these practices to OST. However, other applications are possible, and the committee prefers to leave to OST management the task of translating how these practices could best be adapted and implemented. For example, "hire the best people possible and maintain expertise" would refer in private industry to the cadre of researchers within the company; for OST, which is a collection of program units run by program managers who contract out the technology development work to PIs, the application of this concept might be best translated as "select the most competent contractors."

ROLE AND IMPORTANCE OF EFFECTIVE REVIEWS

The fourth and final explicit issue of the statement of task is "the role and importance of effective peer reviews in the decision-making process." Peer reviews—that is, reviews by technical experts who are independent of and external to the program of work being reviewed—are a vital part of a credible decision-making process. Recent NRC reports (1997b, 1998b) offer guidance with respect to the peer review of individual technology projects as they progress through development stages. As noted previously, the committee recommends the use of peer review in the selection of a new project for funding and at "gate" reviews that serve as later decision points to determine whether to continue funding that project's development.

In addition, the committee considered other key decision points where independent, external review could be valuable, and identified two such opportunities: (1) the point at which OST headquarters managers develop budget targets for major OST program units, and (2) the point at which OST program unit managers establish their statements of technology needs. *Recommendation on (1)*: **OST should have an independent, external body review the bases for the annual decisions, made at the OST upper management level, that establish budget targets for OST program units.** An existing organization such as the Environmental Management Advisory Board (EMAB), composed of representatives of interested and affected parties, could be structured to carry out such a review. **The recommendation on (2) is, as discussed previously, for OST to have a role (e.g., through its contractors) in reviewing site remediation functional flowsheets.** These two types of proposed reviews can be structured to have some, but not all, of the requisite attributes of a peer review (i.e., to be conducted by independent, external, technical reviewers); hence they might be described as "program" or "policy" reviews (for issue (1)) or "quasi-external" reviews (for issue (2)).

PROGRAM CHALLENGES AND MEASURES OF SUCCESS

This section discusses the way in which the committee's recommendations address current issues, such as how to measure success, that challenge the OST program. OST's chief program goal is the use (or deployment) of OST-developed technologies in DOE-EM applications, either to reduce costs, technical risks, and/or schedules of waste management or cleanup activities or to perform a technically difficult or intractable task. Hence, the OST processes (or programs) could be measured by their outputs, as represented in measures such as the following:

- the sum of the cost impacts of each deployed technology;
- the estimates of risk and/or schedule reductions achieved by the use of OST technologies; and
- the enabling nature of OST-developed technologies to assist in difficult and/or intractable tasks.

Other guidance on the generation of appropriate measures and the evaluation of a program against them can be found in NRC (1999b).

Deployment of OST Technologies

Thus far, OST has had only a modest percentage (approximately 20 percent) of its technology projects deployed at DOE sites, in part because of conditions, or "barriers," outside OST's control. For example, DOE site managers are under no obligation to use the results of OST technology development activities. However, OST's deployment record is also due in part to the way it operated in the past, when technologies were often developed without due consideration to input from the site problem owners. In more recent times, these site inputs have been obtained through the STCGs.

Upper-level OST management has decided to address this "deployment barrier" explicitly with programs such as the ASTD program, in which OST provides funding for the first on-site deployment of a fully demonstrated technology to a site that can show the potential for multiple uses of the technology in DOE-EM and an associated cost benefit. In 1997, OST received many more site proposals under these terms than could be funded, as an indication that DOE-EM sites can be given sufficient financial incentive to introduce fully demonstrated technology into their plans to expedite cleanups and accomplish remediation goals.

However, the deployment of OST-developed technologies is not in general subject to these same site incentives, particularly if an OST-developed technology lacks full approval and wide acceptance as a proven technique, or if it cannot be integrated into the site baseline approach. The degree to which sites are unwilling and/or unable to assume some risk, cost, and/or delay in (1) substituting an OST-developed technology for a more familiar baseline technology or in (2) changing the baseline flowsheet approach are two examples of additional disincentives that are probably significant but that the ASTD program does not measure. The conclusion is that the ASTD program does not address all the factors regarding the adoption of new technology that contribute to the OST deployment barrier.

In the committee's analysis, OST's deployment barrier is a symptom of the challenges imposed by organizational structure. Both the inputs (technology needs) and the outputs (deployments) of OST's decision process are controlled by the other DOE-EM offices that are the users of the technology; hence deployment success requires coordination across program boundaries. The OST approach of performing RD&D in a centralized program office (e.g., OST) has the potential advantage of avoiding duplicative developments at multiple sites and coordinating development work across all of DOE-EM. However, for centralized RD&D to be successful within DOE-EM, as determined by deployment of OST-developed technologies or other measures, OST must (1) have good and current information on user technology needs, (2) develop technology products of interest to the user community, and (3) cultivate a customer within DOE-EM for these products that are ready for demonstration and deployment. All of these steps require interaction and cooperation between OST and other EM program offices.

Two previous recommendations—on the identification of needs and top-level strategic goals—propose efforts that both OST and the other DOE-EM offices should undertake to promote technology deployments. **As noted above, OST should identify technology needs based on (1) close and continuing interaction with knowledgeable site personnel and (2) vulnerabilities (i.e., process steps having significant probability of technical failure or underperformance) in baseline remediation functional flowsheets.** Deployment will continue to be a challenging barrier if the different DOE-EM program offices have different goals and incentives related to technology development. **As noted above, OST should establish top-level strategic goals in order to guide internal RD&D efforts to be aligned with the priorities of other DOE-EM offices.**

CONCLUDING PERSPECTIVE

With current uncertainties in costs and technical risks in the DOE-EM program,[6] the RD&D function is necessary. The committee makes no finding of other possible organizational structures (i.e., different from the current DOE-EM program boundaries and delineation of responsibilities) that could be devised to accomplish the RD&D needed in DOE-EM. Statements in this report are thus specific to the context and associated challenges provided by OST's RD&D activities undertaken to expedite the site cleanups conducted by the other DOE-EM offices.

---

[6]DOE-EM cleanup activities are estimated to represent in excess of $100 billion dollars over many decades. OST accounts for a relatively small part of this total DOE-EM budget; OST funding from FY 1991 to FY 1998 was approximately $2.8 billion, or 6 percent of the DOE-EM $45 billion budget during that period.

1

Introduction

Decisions that any technology-based organization makes for research, development, and demonstration (RD&D) result in allocations of resources to achieve certain ends. The decision-making process is the ongoing multiparty conversation that leads to those allocations.

Better decision making can contribute to better use of an organization's resources in many ways, including better coupling to "customer" needs, better selection of high-productivity options, faster achievement of results, and more successful deployment of results. The quality of decisions can always be refined, with improvements sought at various conceptual levels: philosophical, process, and procedural. To do this for the decision-making process, like other management processes, it is necessary to define the process, the players, and the steps involved.

Decision processes provide structure to the evaluation that is needed when

1. desired results or goals are defined,
2. alternatives to meet these goals are identified,
3. relevant information on alternatives is collected and made available, and
4. values are applied, as in evaluations using prioritization criteria or practical experience, to select among the alternatives (see March, 1999).

A successful decision process results in the development of decisions and the execution of efforts that support an organization's goals.

Decision making in simplified situations (e.g., bounded by clearly defined goals, limited alternatives, and limited access to applicable information) may not require much formality or structure, in which case informal decisions (such as those made by an individual's subjective judgment) may be adequate. However, the degree and frequency of success for informal decision making diminishes sharply the further the circumstances depart from ideal or simplistic situations. Large organizations such as government agencies customarily deal with complex situations, uncertainties, long time horizons, and a range of stakeholder[1] concerns. Such organizations would therefore stand to benefit from sound, well-defined, structured decision-making processes that

- direct the collection and use of information,
- allow for transparency and stakeholder access,
- permit the use of metrics to assess and redirect efforts over time,

---

[1]A "stakeholder" as used here is anyone with a stake in the program or process involved, that is, an "interested and affected party" (NRC, 1996d). For DOE-EM, the U.S. taxpayer is in principle a stakeholder, but the term is commonly used to denote constituencies local to any of the DOE-EM sites.

- provide the opportunity to conduct retrospective analyses to elucidate and apply any "lessons learned," and
- address alternatives as a hedge against failure.

The complexity of the Department of Energy's (DOE's) RD&D decisions and the context in which these decisions are made warrants the use of formal procedures. However, some managerial flexibility is warranted in funding decisions, particularly in response to late-breaking development opportunities, since the nominal Office of Science and Technology (OST) procedures of assessing technology needs and initiating suitable development projects are managed on an annual time frame to correspond with federal budget provisions. The DOE's Environmental Management (DOE-EM) program has RD&D needs—to address radioactive waste management activities, the long-term disposition and dismantlement of nuclear facilities with no further mission, and the environmental cleanup of contaminated DOE sites. The Office of Science and Technology within DOE-EM funds RD&D projects on technologies having application to these waste management and remediation challenges. The complexity of OST decision making stems in part from the diversity of technical areas and in part from its institutional environment and other considerations discussed later in this report.

PURPOSE OF THIS REPORT

This report examines the decision-making processes of the Office of Science and Technology within the DOE-EM program. Several National Research Council (NRC) reports have identified issues associated with the DOE-EM program and with OST in particular (see Appendix A).[2] The nature and quality of the decisions made in allocating resources to RD&D activities is a topic that has been addressed in those reports. At the request of DOE, the NRC undertook the present study "to evaluate the effectiveness of the OST decision-making process and make specific recommendations to improve it, if appropriate" (see statement of task in Box ES.1). As is customary in NRC studies, committee members did not represent the views of their institutions but formed an independent body to author this report. The purpose, scope, and activities of OST and DOE-EM are discussed briefly below.

DOE-EM AND OST

The DOE's Office of Environmental Management was established in November 1989 to address environmental contamination and other cleanup problems resulting from nearly 50 years of nuclear weapons production, reactor fuel and target rod processing, and RD&D activities at DOE sites. One of several program offices within EM,[3] the Office of Science and Technology (commonly referred to by its "mail stop"

---

[2]Appendix A summarizes the findings and recommendations of past NRC reports that relate to decision making within DOE-EM. They are mentioned here to show the statements that have already been made on the subject. In general, this report is in agreement with the essential points made in the earlier reports and is a natural extension of them. The findings and recommendations offered in this report are intended to augment them by a more in-depth examination of OST than these earlier studies had the opportunity to conduct.

[3]Other separate EM program offices are the Office of Waste Management (EM-30), the Office of Environmental Restoration (EM-40), and the Office of Nuclear Material and Facility Stabilization (EM-60).

as EM-50 and known prior to 1996 as the Office of Technology Development) supports project work in basic environmental science, risk assessment, technology development, and technology deployment. OST has no environmental liabilities of its own; hence it is not directly subject to compliance requirements or cleanup agreements.

OST has been the object of reviews and scrutiny by several entities interested in its value and accomplishments (GAO, 1992; 1994; 1996; 1998; Holt and Day, 1997; NRC, 1996b; Rezendes, 1997; Surles, 1997). In particular, questions have been raised about how to assess the OST program with respect to its return on investment and whether the return is worth the program's cost. The answer to these questions resides in the nature and magnitude of the contributions made (and a projection of contributions that will be made) by OST's technology developments to actual site cleanup and other DOE operations. Analysis of OST's performance requires that records be kept of expenditures made and technologies deployed at sites and that some credible approach be devised for evaluating the actual and potential contributions of OST's technologies.

Because technologies are often slow to mature and their applications may come about over many years, meaningful present-day evaluations of their worth are difficult to make. Nonetheless, at least some products of a truly worthwhile technology development program should have some acceptance and impact on DOE-EM site cleanups in the relatively short term. If the decision-making processes employed by OST for selecting the waste management and cleanup technologies it supports (and for encouraging their development and deployment) are sound and are embedded in a sound framework of good decisions by its parent organization, then OST can play a valuable role in site cleanup.

OST PROGRAM UNITS FOR RESEARCH, DEVELOPMENT, AND DEMONSTRATION

Since 1989, OST has selected, funded, and managed RD&D projects whose successful completion and deployment should result in improvements to the technologies underpinning DOE-EM cleanup work. These activities have been carried out in several OST program units: four Focus Areas; three Crosscutting Programs; an Industry Program; a University Program; a Technology Integration Systems Application (TISA) International Program; and an Environmental Management Science Program (EMSP) (see Table 1.1).

The names of the four Focus Area programs (Subsurface Contaminants, Mixed Waste, Decontamination and Decommissioning, and Tanks) and the three Crosscutting Programs (Efficient Separations and Processing; Robotics; and Characterization, Monitoring, and Sensor Technology) represent seven major categories in which technology needs and corresponding development work exist. OST managers in each of these program units make decisions, based on an assessment of priorities, to initiate and to continue (or terminate) funding technology development projects. The result of these decisions is a suite of technology development projects intended to address DOE-EM cleanup and waste management problems.

In addition to the Focus Areas and Crosscutting Programs, OST inaugurated the Industry Program to enhance technology input from private industry; the University Program to establish research and development centers at a few select universities; and the TISA International Program to involve foreign cleanup expertise, especially that of the former Soviet Union. In 1995, the EMSP was mandated by Congress to ensure that a fundamental knowledge base would be developed and maintained from which technologies could be generated to meet the longer-range DOE site cleanup problems for which there may not now be efficient, effective, economical, and/or acceptable solutions.

TABLE 1.1   Major OST Program Units in FY 1997-1998

OST Program Unit	Management Location
Focus Areas	
Decontamination and Decommissioning Focus Area	Federal Energy Technology Center (FETC) Morgantown, West Virginia
Tanks Focus Area	Richland Operations Office Richland, Washington
Subsurface Contaminants Focus Area	Savannah River Operations Office Aiken, South Carolina
Mixed Waste Focus Area	Idaho Operations Office Idaho Falls, Idaho
Crosscutting Programs	
Robotics Crosscutting Program	Albuquerque Operations Office Albuquerque, New Mexico
Efficient Separations and Processing (ESP) Crosscutting Program	Oak Ridge Operations Office Oak Ridge, Tennessee
Characterization, Monitoring, and Sensor Technology (CMST) Crosscutting Program	Nevada Operations Office Las Vegas, Nevada
Industry Programs	Federal Energy Technology Center (FETC) Morgantown, West Virginia
University Programs	Federal Energy Technology Center (FETC) Morgantown, West Virginia
Technology Integration Systems Application (TISA)	
TISA Domestic Program	DOE Headquarters Washington, D.C.
TISA International Program	DOE Headquarters Washington, D.C.
Other OST Programs and Groups	
Environmental Management Science Program (EMSP)	DOE Headquarters Washington, D.C.
Accelerated Site Technology Deployment (ASTD, formerly Technology Deployment Initiative (TDI)) Program	DOE Headquarters Washington, D.C., and Idaho National Engineering and Environment Laboratory, Idaho Falls, Idaho
Technology Management System (TMS)	DOE Headquarters Washington, D.C.
Site Technology Coordination Groups (STCGs)	Each major DOE-EM site

Expansion of OST's Role in Deployment

In addition to selecting, funding, and managing RD&D projects, OST supports work to facilitate deployment of the resulting technologies on DOE-EM problems.[4] Because OST performs RD&D activities on behalf of (and to be deployed by) other DOE-EM program offices, the deployment of OST-developed technology is in part a function of the interactions of these program offices. OST has learned that it cannot stop with the development or even the demonstration of a technology and expect that the technology will be used by the problem owners at the sites. There exists a need for additional action to encourage the site "end user" (i.e., the combination of the DOE site manager with responsibility for the problem for which the technology is intended and the site contractor responsible for implementing a remediation approach) to adopt and deploy any new technology. Consequently, OST has proposed and, with congressional approval, implemented several programs whose scope appears to be outside a narrow interpretation of its original charter (U.S. Congress, 1989) of engaging in technology development, to perform this "deployment facilitation" function. In general, these programs were proposed to facilitate OST interactions with the technology user to engender the user's deployment of OST's technologies and those available from other sources. Such deployments must occur if OST is to play a useful role by introducing efficient, effective; acceptable, and/or economical new technologies into site cleanups. Examples of programs initiated by OST to increase technology deployment include the following:

- The TISA Domestic Program, intended to facilitate regulatory and stakeholder knowledge, communication, and acceptance of new technology applied to DOE-EM problems;
- The Large-Scale Demonstration Projects (LSDPs) of the Decontamination and Decommissioning Focus Area (DDFA), intended to bring industrial technologies into DOE-EM for use on decontamination and decommissioning (D&D) applications; and
- The Accelerated Site Technology Deployment (ASTD, formerly the Technology Deployment Initiative, or TDI) Program, intended to facilitate technology deployment.

The ASTD program is an especially noteworthy result of OST's determination to have improved, innovative technologies deployed at the sites. The mission of the ASTD program is to cofund DOE-EM site proposals to use already-demonstrated technologies to expedite site cleanups, in situations where the intended deployment would achieve a significant cost savings. ASTD was not directed at developing new or improved technologies, but rather was designed to overcome barriers to the use of technologies new to DOE-EM. The appendixes to this report provide further description of the ASTD and TISA Domestic Program (Appendix E) and the LSDPs of the DDFA (Appendix C).

Tracking the Benefits of OST Technology Development Activities

In an attempt to monitor its programs better, and thus gain insight into the validity of its decision-making practices, as well as demonstrate that it was doing a worthwhile job of developing and implementing new technologies, in 1997 OST put into place the Technology Management System (TMS), a database to keep track of the technologies it developed, their costs and cost savings, and their deployments. This system

---

[4] Although deployment of OST-developed technology outside DOE-EM is possible, the deployments within DOE-EM provide accomplishments that are directly related to the OST and DOE-EM missions.

provides a tool for OST to use in documenting, examining, and evaluating its performance. The number of deployments is an important measure of OST performance that is used by Congress and other interested parties, and reflects in part on the quality of decisions that have been made.

An additional step taken by OST was to employ the U.S. Army Corps of Engineers to prepare estimates of cost savings over baseline process costs (DOE, 1997j; U.S. Army Corps of Engineers, 1997). A serious difficulty with this approach is in obtaining good estimates of baseline costs (NRC, 1998a). However, relative costs—that is, differences in cost between competing technologies or processes—may be estimated more reliably than absolute costs of total projects and are of value in providing direction on which technologies to support.

IMPORTANCE OF BUY-IN BY TECHNOLOGY USERS

OST's decisions on which technologies to develop and attempt to implement at the sites is strongly affected by influences originating outside OST. Two primary external influences are (1) the budget it receives from Congress and (2) the problem owners, or site managers—who are customers who both define the technology needs provided to OST and use the technologies from OST. These two external influences are not independent of each other. Failure of the site problem owners to use the technologies can make OST's efforts appear to be without value, which could adversely affect congressional funding actions. Consequently, OST must fund projects that are needed by the problem owners at the sites, who must be convinced of the merits of the technologies and must use them. This must be done often in the face of reluctance on the part of the sites to make changes to a baseline process that might result in inconvenience, liability (financial or professional), delay, adverse stakeholder reaction, or problems with regulatory agencies. There also may be a desire by the problem owner to receive RD&D funds directly rather than have them go to another party, such as OST, particularly if the latter is perceived as not being as responsive to problem owner needs as the problem-owning office would be if it received the funds directly.

Although OST's operating environment is different in many ways from that of most industries, there are good industrial decision-making practices for technology RD&D that are helpful in achieving problem owner buy-in. One such practice is the strong emphasis in industry on customer involvement in decisions on technologies to develop. This involvement works in both directions: the industrial RD&D organization obtains input from the customer on needs, and the customer learns from the RD&D organization about technologies and the industry's technology development capabilities.

In the early 1990s, OST did not fully recognize the necessity of obtaining customer (problem owner) buy-in. However, OST is now acutely aware of the importance of paying close attention to customer needs and, in 1994, put into place mechanisms for improved interactions with site personnel regarding site needs. For example, a program reorganization in 1994 created the current user-driven approach to identifying technology needs as the basis for investment in technology development projects. Additional OST efforts to interact with the end customer and to overcome resistance to deployment of its technologies are discussed in Chapter 2.

REPORT SCOPE AND ORGANIZATION

The approach taken by the committee for this study was first to gather information[5] by visiting the DOE-EM sites where the major OST program units are managed (see Table 1.1 for a listing of the program units). The prioritization and decision-making processes of each of these major OST program units are described in Appendixes B-E, summarized in Chapter 4, and evaluated in Chapter 5. After the site visits, the committee

- developed an historical overview of DOE-EM and OST and the evolution of OST to its present organizational structure;
- investigated how some large private industries and non-profit institutions make technology RD&D decisions; and
- developed a decision-making model (see Figure 4.1) that embodies the essence of the context in which OST makes its decisions but is, in fact, a generic model for technology RD&D in the federal government environment.

Chapter 2 presents an historical overview of the evolution of DOE-EM and OST in response to a recognition of the ineffectiveness of past practices for getting technologies deployed. This evolutionary development led to the current organizational structure and decision-making practices, which are themselves being refined in a continuing evolutionary effort to improve them.

Chapter 3 and Appendix F identify relevant "best practices" by industry in technology development decision making. To suggest how OST might improve its prioritization and decision-making processes, the committee considered the decision-making practices of some other organizations, including many of the leading RD&D companies in private industry. This was largely done using the results of a multiyear, multicompany RD&D decision quality benchmarking study carried out by Strategic Decisions Group (SDG), the RD&D Decision Quality Association, and the Industrial Research Institute (see Appendix F). The committee also interviewed five companies with substantial environmental remediation programs. In addition, site visits were made to the Gas Research Institute (GRI) and the Electric Power Research Institute (EPRI), institutions that face many of the same organizational complexities as DOE, to study their RD&D project selection and prioritization processes. Summaries of information acquired at three of these interviews are contained in Appendixes G-I. Finally, the committee drew upon published literature describing the use of decision and risk analysis, probabilistic risk analysis, and cost-benefit analysis to help make similar RD&D decisions in a wide variety of settings.

Chapter 4 presents a model process for technology decision making in the federal government environment, addressing important constraints such as the congressional budgeting process and the interactions among OST, other DOE-EM entities, and the contracted site operators.

Chapter 5 contains discussion supporting the committee's findings and recommendations, which are presented in Chapter 6, on the nature, appropriateness, and effectiveness of OST's decision-making processes for selecting and funding RD&D activities.

This chapter concludes with a discussion of the way the subject of decision making, particularly in RD&D contexts, is formulated, analyzed, and practiced.

---

[5]The committee concluded its information gathering from DOE sources in April 1998.

DECISION MAKING AS A DISCIPLINE

Structured decision making involves setting goals, defining alternatives to achieve these goals, and collecting information about the likely consequences of each alternative (March, 1999; and Zeleny, 1982). In many real-world situations, relatively simple circumstances often apply. For DOE-EM, these situations may occur, for example, after a given development project scope is defined, goals are agreed upon, and measurement milestones and funding are set and well documented. Then the further set of decisions involved in executing and monitoring .the project is straightforward—provided the goals or measurement milestones do not change.

In many decision-making contexts, and in most DOE-EM situations, these simple circumstances rarely prevail. As expounded in many business courses and books, it is no longer practical for a business organization to have a single dominant and unchanging goal (such as bottom line return) or a simple vision (such as making the best "widget" of its type). The goals and missions themselves must be subject to continuing critical review, explicit articulation, and refinement (e.g., by a change in regulations). In the public sector, this kind of review can be more complex because of the larger number of stakeholders often involved and the existence of multiple, potentially conflicting goals.

In addition to changing goals and missions, other refinements over time can occur in the criteria used to select and evaluate RD&D proposals, the relative weighting of these criteria, and the information available for evaluation against these criteria. In industry, a basic criterion such as the relative cost-benefit of several options commonly serves to screen out weak items. In government RD&D, the simplification of having one or two dominant factors is rare; many relevant factors are often evident, and commonly more than two of them will appear to have comparable importance or weighting. Moreover, the information on some choices may be fragmentary or evidently unreliable, but this may be a reason for doing some exploratory RD&D.

Development and Growth of Decision Disciplines

The evident deficiencies—in business, military strategies, government regulation, and legislation—of oversimplified approaches to complex decisions has contributed to the development of the discipline of operations research, which involves modeling the system or the situation in order to recognize a framework for decisions. As part of this development, a scientific and engineering approach to a formal, prescriptive theory of decision making has also been created. Seminal books and publications (e.g., von Neumann and Morgenstern, 1947; Edwards, 1954; Luce and Raiffa, 1958; Chernoff and Moses, 1959; Fishburn 1964; Howard 1966;[6] 1983; Pratt, Raiffa, and Schlaifer, 1965; Raiffa, 1968; Tribus, 1969) set the foundations of structured decision making.

More modern references (e.g., Cooper, 1993; Hammond, Keeney, and Raiffa, 1999; Keeney, 1992; March, 1994; 1999; Martino, 1995; Matheson and Matheson, 1998; Saaty, 1994; 1996) also describe decision-making practices. Some references (Simon, Smithburg, and Thompson, 1991; Simon, 1997) connect decision making to other closely related disciplines or to other components and issues (e.g., organizational structure, administration and policy, and efforts to promote communication and teamwork between different units) within an organization. The committee's approach in conducting this study was

---

[6]This is the first article to use the term "decision analysis."

not to produce a report similar in character to these treatments. Instead, the committee sought to identify the important decisions facing the sponsoring program office (OST), its current decision-making processes, and the important influences on these processes. The result is a series of findings and recommendations intended as general guidance on decision making for OST (or any organization in a similar context), given its important constraining features.

The description and consideration of OST decision processes were difficult to separate from descriptions and considerations of organizational structure, institutional procedures, and program management. Indeed, the committee had to consider these and other program elements because of their relevance to the prioritization and decision processes. Therefore, some report language, although relevant to decision making, is descriptive of and applicable to these other programmatic elements. These comments are offered only insofar as they impact decision-making processes, which the committee was obliged to consider (see statement of task in Box ES.1).

Risk Aspects of RD&D Decisions

A closely related development, probabilistic risk analysis (PRA), has occurred in parallel to the growth of methods and applications of decision analysis. The logic, data, and arithmetic processes involved in the disciplines of risk assessment and decision analysis have close similarities. Both disciplines involve realistic modeling and require systematic treatment of the inherent uncertainties in the information and data available. Risk evaluations commonly are among the most heavily weighted factors in many kinds of decisions.

For environmental RD&D, three kinds of risk evaluation are commonly considered. The first is the probability of success or failure of the RD&D to meet its intended performance goals (i.e., a risk of failure to meet a technical specification). The second is the extent to which the expected result of the RD&D can serve to reduce significant environmental hazards to workers and the public (i.e., a risk to the environment or to people). The third is the extent to which a successful RD&D result will be used in the field (i.e., a risk of failure to be deployed). Even if the work is technically sound and solves an important problem, it may fail to get much use for reasons unrelated to the technical quality of the RD&D output. The context in which "risk" is used in this report makes clear which of these three meanings is intended.

2

Historical Evolution of OST Decision Making

This chapter provides a brief discussion of the evolution of the DOE-EM program and of OST in particular. OST's decision-making structures have evolved over time, with an increasing responsiveness to site user-determined needs and a greater focus on monitoring and assisting the deployment of technologies for DOE-EM site cleanup and waste management activities.

BACKGROUND OF DOE-EM AND OST

When the Manhattan Project began, DOE and its predecessor agencies were not subject to external federal or state regulation. Beginning in the 1970s, however, federal legislation waiving the federal government's sovereign immunity from state and federal environmental laws was enacted. Nevertheless, DOE resisted application of external chemical hazardous waste regulations for several years, until a suit (*LEAF v. Hodell*) was successfully brought against it. A 1984 federal court decision rejected DOE's contention that the 1976 Resource Conservation and Recovery Act (RCRA)—the nation's centerpiece of "cradle-to-grave" hazardous waste management legislation—did not apply to DOE because of the Atomic Energy Act. In 1987, DOE acknowledged that RCRA applied to the hazardous waste component of mixed (i.e., hazardous and radioactive) wastes. The need for DOE to comply quickly with RCRA was made more compelling by the 1992 Federal Facility Compliance Act, which specified that federal agencies would be subject to fines and penalties for violations of RCRA. In addition, section 120 of the 1986 Superfund Amendments and Re-authorization Act made clear that the 1980 Comprehensive Environmental Response, Compensation, and Liability Act (CERCLA, or "Superfund" law) applied to federal facilities, giving the U.S. Environmental Protection Agency (EPA) and the states explicit oversight responsibilities concerning the cleanup of legacy waste sites.

Early Years

The first Assistant Secretary of DOE-EM, Leo Duffy, serving under then Secretary of Energy Admiral James Watkins, approached the mission of EM in the optimistic "can-do" spirit that had typified the nuclear weapons research, development, and production missions of DOE and its predecessor agencies, as well as the early optimism that the U.S. Congress and others had brought to the task of cleaning up the nation's legacy waste sites under Superfund. This attitude was reflected in commitments made by DOE, beginning in 1989, within the context of triparty agreements (DOE, EPA, and the host state) concerning remediation at several major DOE sites. By the early 1990s, however, this optimism had become tempered by the realization that the cleanup of DOE sites was likely to take far longer and require far more funds than first anticipated (Russell et al., 1991).

The early years of OST have been characterized to the committee as a time of headquarters control over decisions (Frank, 1997b). During these inaugural years, program staff were hired and money was allocated directly by then Deputy Assistant Secretary for Technology Development Clyde Frank, prior to the establishment of a more formal program structure for decision making.

Integrated Demonstrations and Integrated Programs

In a reorganization of the program structure of EM-50 in the early 1990s, several program units (i.e., integrated demonstrations and integrated programs) were created. Some of these were administered at field sites, with an emphasis on large-scale demonstration of developed technologies as a key goal.

In setting up this technology development program structure, DOE proceeded on the basis that currently available, conventional cleanup and waste management technologies were not always effective, and could be costly, and that new technologies were needed to remedy these shortcomings. As a result, DOE initially established three major RD&D areas for the technology development program—ground water and soil cleanup, waste retrieval and processing, and waste minimization and avoidance. Each major RD&D area was to be supported with "integrated demonstrations" in which multiple technologies would be tested at a particular DOE site and would ultimately result in a complete system to address a site problem, from technologies for site characterization and remediation to technologies for monitoring (GAO, 1992).

The first integrated demonstration of technologies for cleaning up chlorinated solvents in soil and ground water at the Savannah River Site was initiated in 1990 and was followed by seven additional integrated demonstration projects begun in 1991. By 1992, however, DOE officials realized that the integrated demonstration approach required more resources than were available and began to revise their approach toward one in which individual technologies, rather than integrated technology systems, would be emphasized. The individual technologies were to be focused in specified areas, such as characterization and monitoring and mixed waste processing (GAO, 1992). By 1994, approximately 10 integrated demonstrations, four integrated programs, and seven other program units within OST had been formed (DOE, 1994b).

Despite the efforts in creating these programs and demonstrations, few of the new technologies were actually used in cleanup applications. These issues helped shape the next internal DOE reorganization and new program management structure.

The new structure was intended to rectify several problems. Some of them had been identified in a report that noted the technology development program's lack of measurable performance goals, clear cost estimates, and schedules for integrated demonstration projects (GAO, 1992). This report also noted the absence of clear decision points to eliminate poorly performing or inappropriate projects.

Also identified by DOE (DOE, 1994a) and noted in a subsequent GAO report (GAO, 1994) were other problems: a reluctance on the part of various groups (including regulators, local officials, and other stakeholders, as well as field officials and on-site contractors) to endorse the use of innovative technologies; failure of DOE's program offices, especially the Offices of Waste Management (EM-30), Environmental Restoration (EM-40), and Science and Technology (EM-50), to work together effectively; and flawed decision making, which excluded EM-50's technical experts from contributing to technology decisions at particular sites. To remedy these problems, new OST program units were created in 1994, as described next.

New Headquarters Priorities and New OST Program Units, 1993-1994

In 1993, under the Clinton administration, Hazel O'Leary was appointed Secretary of Energy, and in May 1993, Thomas Grumbly became Assistant Secretary for EM. He brought to his DOE position a philosophy of openness and "stakeholder involvement," as did Secretary O'Leary (DOE, 1994a). In addition, under Grumbly a new emphasis was placed on *risk* as the most important criterion for deciding which aspects of cleanup should be tackled first, as well as a focus on technology development and strengthening relationships with the private sector (DOE, 1993b).

In 1993, a working group was established by Grumbly to develop a new approach that would focus environmental research and technology development activities across DOE on its most pressing environmental restoration and waste management problems. Although EM-50 had, since 1989, been responsible for managing DOE's environmental applied research and technology development, substantial related and separately funded technology development activities were being carried out by EM-30 and EM-40, as well as by private industry external to DOE-EM. (As a result of budget pressures, most of the EM-related RD&D in these other DOE organizations ceased as of FY 1998.) In addition, other DOE offices, such as the Office of Energy Research, were seen as potentially contributing to EM technology development efforts.

Grumbly's working group included participants from across DOE and had customer and supplier representation (DOE, 1993a). The group's efforts started in August 1993 and led to the document, *A New Approach to Environmental Research and Technology Development at the U.S. Department of Energy* (DOE, 1994a), which identified five[1] major problem areas, called Focus Areas: contaminant plume containment and remediation; mixed waste characterization, treatment, and disposal; high-level waste tank remediation; landfill stabilization; and facility transitioning, decommissioning, and final disposition. The Focus Areas, and other activities such as crosscutting RD&D areas, were to be managed by a new team structure. Key features of the new management approach (DOE, 1994a) were the following:

- A Steering Committee composed of upper-management representatives from each organization, with other top-level DOE officials serving as ex-officio members. The committee's purpose was to ensure that the technology development program remained responsive to EM priorities.
- Management Teams for each of the Focus Areas, with the core team consisting of one representative each from other major offices within the EM program, as well as "an Operations Office representative, as appropriate" (DOE, 1994a, p. 1-9.)
- Lead Organizations for each of the Focus Areas—the lead organization being selected by the Management Team with the approval of the Steering Committee, based on proposals submitted by candidates. This organization could include "DOE Operations Offices, National Laboratories, Management and Operations (M&O) Contractors, universities, non-profit or not-for-profit organizations, industry, or other interested organizations" (DOE, 1994a, pp. 1-13). Individual organizations or consortia could apply. In practice, management of the Focus Areas and Crosscutting Programs has been established at DOE-EM operations offices.
- Implementation Teams for each Focus Area, supervised by the Lead Organization, to be responsible for day-to-day technical management within the Focus Area.
- Focus Area Review Groups, whose membership was to be chosen on an ad hoc basis from scientists, engineers, and interested stakeholders to work with the Management and Implementation Teams.
- Site Technology Coordination Groups (STCGs) at each DOE-EM site, to work with the Management and Implementation Teams of the various Focus Areas to ensure that site needs were identified and addressed and that technical solutions were implemented. Membership of each STCG was to include Operations Office personnel, together with operating contractor and laboratory personnel from the site. Local regulators were to be kept informed of technology development, and stakeholder and public interest groups were to be actively involved with each STCG.

---

[1] Two of these five, the Contaminant Plume Containment and Remediation Focus Area and the Landfill Stabilization Focus Area, were combined in 1996 to form the Subsurface Contaminants Focus Area.

Much of this program structure was adopted in 1994-1995 and is in place today. The major restructuring of OST's way of doing business in early 1994 was a significant midcourse correction—a shift away from the informal mode of decision making in OST's first years, toward a much more structured, multilayered approach, as well as a shift from developer-driven to user-driven priorities. This shift was made within the context of increasing pressure to show results from a program that had, at that point, spent more than $500 million with few of its technologies deployed at the sites.

Implementation of the New Approach

Focus Areas, Crosscutting Programs, and STCGs were chartered in 1994 (see Box 2.1), although all STCGs were not fully operational until later. During 1994-1995, the lead management of the Focus Areas was moved to DOE field offices, and during 1997, Crosscutting Program management was also moved from headquarters to field offices.

Sharpening the OST Focus on User Needs

Prior to 1995, the DOE-EM cleanup program had no firm total cost estimate associated with it. Technology development activity in the Integrated Demonstrations/Industry Program (ID/IP) was conducted in the spirit of developing and demonstrating good projects, which if shown to be meritorious would presumably lead to their recognition and implementation by the user community.

The year 1995 brought several developments that changed this approach. The "Train Wreck report" (Blush and Heitman, 1995) noted the first major budgetary shortfall for DOE-EM; in response to congressional inquiry into the future and utility of national laboratories, the Galvin report (DOE, 1995e) argued for their continued support, and the Baseline Environmental Management Report (BEMR) (DOE, 1995c) provided the first total life-cycle cost estimate for the 70-year EM program. The year 1995 was the "high-water mark" for OST funding, as well as the beginning of reduced funding for internally funded technology development work conducted by the other, problem-owning EM offices.

The reports cited above, along with budget cuts, helped shape a climate in which OST focused on users' technology needs for the near future and on accomplishing technology development work responsive to such user requests. The STCGs were established in 1994-1996, and needs statements were collected formally and used in program planning, as a key basis for making decisions on what projects to fund.

1996-1997 Changes in Top-Level DOE-EM Goals

In 1996, Thomas Grumbly resigned and Alvin Alm was appointed Assistant Secretary for EM. Soon after assuming this position, Alm introduced a new priority objective of *reducing the mortgage* of the DOE-EM program by completing the cleanup of as many DOE sites as possible within 10 years, an effort supported by Secretary Federico Pena when he became Secretary of Energy in early 1997. The strategic plan to achieve this—the "Ten-Year Plan" (later called the "2006 Plan" and still later renamed the "Accelerating Cleanup: Paths to Closure" plan), has 2006 as the completion goal for most sites, although not for the major DOE-EM sites (DOE, 1998a). In this management approach for the EM program, cleanup jobs were cast as a set of separate projects to be managed, most to be completed within 10 years, with priority given to work that would reduce the program's budgetary requirements beyond the year 2006 (i.e., reduce the mortgage by doing work to place facilities in low-cost maintenance or remediated conditions). The increasing emphasis by OST on meeting near-term needs of cleanup activities scheduled to be completed

BOX 2.1   Summary of OST Program Structure, 1994-1998

The major internal program units within OST that fund technology development projects are Focus Areas and Crosscutting Programs, which are organized around subject areas (see Table 1.1 for these names) reflective of DOE-EM cleanup challenges. Each program obtains input in the form of lists of technology needs from each site. Additional program units, not shown below, address university and industrial technology input and international collaborations.

Site Technology Coordination Groups

Each major EM site has an OST-funded STCG, which is a group comprised in some cases of DOE employees only and in other cases with representation from site contractors and interested non-DOE stakeholders. The STCG is tasked with developing a list of site technology needs based on input from the site problem owners, prioritizing them, and forwarding them to OST Focus Areas. The STCGs generally have working groups in each major subject area. More detail on the composition, function, and methods of STCGs is presented in Appendix B.

Focus Areas and Crosscutting Programs

Focus Areas and Crosscutting Programs are OST program units (see Table 1.1 and Appendixes C and D) that perform the following tasks:

- compile site-specific technology needs into a single DOE complex-wide list;
- prioritize these needs to reflect complex-wide priorities across sites;
- generate packages of work as budget entities for which to solicit funding;
- solicit technology development work proposals by issuing Requests for Proposals and evaluating the responses;
- allocate funds to projects selected from among the proposals;
- monitor progress of each funded project;
- demonstrate, when appropriate, the technology at a DOE-EM site, to engender adoption by the user community; and
- involve, where appropriate, a private-sector company in development work.

Each Focus Area and Crosscutting Program has developed a different process and methodology as well as different criteria (see Chapters 4-5 and Appendixes C-D for more detail on these steps).

Other OST Programs

Other OST programs that fund technology development work (see Appendix E) include the Industry Program and the University Program. These program managers interact with Focus Area and Crosscutting Program managers to define relevant and appropriate scopes of work for their contracts, grants, and cooperative agreements. As with the Crosscutting Programs, the Industry and University Programs use the Focus Areas' prioritized lists of technology needs to determine priorities and work scopes. An "iterative-collaborative" process of communication among program managers is used to lead to refinements of the technical specifications for bids and better knowledge by Industry and University Program managers of EM-specific technical challenges that the solicitation is designed to address.

Headquarters Oversight Role

Headquarters oversight of these program units includes the important task of proposing to the Office of Management and Budget and to Congress the division of the overall OST budget among them. The most direct control that headquarters-based OST managers presently exercise is in the formation of targeted budget levels for each OST program unit (with the internal review budget [IRB] planning exercise) and in the refinement of these levels in budget adjustments, such as at the annual Program Execution Guidance (PEG) review. In their overview role, OST headquarters managers are free to look for weaknesses in a proposed program on any basis (e.g., programmatic or technical). Currently the Environmental Management Advisory Board and upper-level management bodies provide some evaluations and input into headquarters decisions.

During 1996-1997, Dr. Clyde Frank, the Deputy Assistant Secretary for OST, created a "Board of Directors," consisting of himself and the most senior DOE official at each of the four DOE sites hosting the Focus Areas. This five-person body made decisions on any aspect of OST's program that it though important to discuss and vote on. Issues decided by such voting included allocation of the OST budget to subordinate program units (e.g., Focus Areas and Crosscutting Programs) within OST (these allocations then became OST's budget requests for those program units, submitted as part of the federal budget process). This board was replaced by the Technology Acceleration Committee in 1997-1998, and, more recently, by the EM Integration Executive Committee.

---

prior to 2006 is provided by "linkage tables" that relate technology needs statements, OST's technology development projects, and the project number (the Project Baseline Summary, or PBS) of the EM cleanup job to which it would be applied.

THE DEPLOYMENT BARRIER

From the beginning, OST faced difficulty in getting its technology developments used in the field. Approximately 700 technologies have received at least partial OST support since 1989; fewer than 20 percent of them have been implemented at DOE-EM sites (GAO, 1998).

Many factors affect the value of this deployment rate. For example, an RD&D program heavily focused on late-stage developments would show a higher deployment rate than a program with more early-stage research-oriented projects. However, an RD&D portfolio overly concentrated on late-stage work would not be in the nation's best interest because it would fund proportionately less innovative work that is at early stages of development. Therefore, a balance must be struck between a significant number of deployments (a necessary measure) and other considerations. One possible way to provide this balance is to add other performance measures, such as estimates of the dollar value of cost and risk reductions that implementations of OST-developed technologies will achieve in the long term.

In response to this perceived technology deployment "barrier," OST made changes designed to increase adoption by the user community of technologies it developed or identified. The ASTD program, formerly the Technology Development Initiative (TDI), was instituted in 1997 to address institutional and other barriers to the deployment of new technologies on DOE-EM site cleanups. Top management level initiatives, including the creation of a Technology Acceleration Committee (TAC), were also designed to address these barriers. These efforts are described in greater detail below.

Accelerated Site Technology Deployment

In 1997, the issue of the worth of the OST program (Holt and Day, 1997; Rezendes, 1997; U.S. Congress, 1997) was raised in the context of a financial return on investment (ROI) to the federal government on the $2.6 billion expended by OST since the start of the program in 1989. In response to attention to this issue, OST created the ASTD program as an EM-50 competitive solicitation to DOE field offices and sites for proposals of how they would use EM-50 funds to reduce a site's cleanup costs. The approach used by ASTD is to underwrite the first-time use of new (new to DOE, but not necessarily new to industry) technical approaches applied to cleanup projects already scheduled to be performed in the near term. A requirement of the proposals was to perform an ROI calculation showing the cost savings (or cost avoidance) relative to the remediation alternative (the baseline method) that would be used in the absence of EM-50 funds. This requirement implied that competitive proposals would go only to those cleanup jobs for which a baseline remediation method already existed.

The ASTD program was not designed to be a demonstration program for unproven technologies. Site proposals were screened to consider only those with fully demonstrated, industrially available technologies that could be used by sites without further development. Proposals were also screened to favor those proposing second and third applications of the technology at other DOE-EM sites, applications that would not be funded by OST dollars.

Implicit in the ASTD approach is the existence of a basis for estimating baseline process costs. The technical methods under comparison were to be technologies with sufficient engineering and performance data for use as a basis for cost estimates. The better-than-baseline technologies could come from any source, either from outside the DOE-EM program or from formerly funded DOE-EM technology development projects that had matured into viable cleanup techniques, since the focus of the ASTD program was to overcome barriers to the adoption of any new technologies in the EM cleanup effort. Of subordinate interest to OST program managers was the degree to which technical innovations that enjoyed past funding support by EM-50 were represented in the suite of new and cost-saving technologies that were part of successful bids.

This program, funded by Congress at $27 million for FY 1998, showed a determined attempt by OST program management to address the internal barriers to implementing new technologies in DOE-EM site cleanups. OST also noted that a large backlog of innovative but maturing technologies exists under OST programs that would benefit from the incentives provided by the ASTD.

With the introduction of the ASTD, OST expanded its mission beyond technology development to promoting the deployment of mature technologies.

Top-Level Efforts to Promote Deployment

A July 3, 1997, DOE memorandum (Alm, 1997) from Assistant Secretary for Environmental Management Alvin Alm called for the creation of an upper-level management committee, the TAC, chaired by Alvin Alm and consisting of the EM Deputy Assistant Secretaries (DASs) and five DOE field office managers. This group in effect replaced the Board of Directors that Clyde Frank had established as the most senior DOE body in which OST has representation. The TAC's chief function was to help to deploy technologies; to this end, the Alm memo called for senior managers to be accountable for achieving results. The TAC met twice (September 1997 and January 1998) to seek ways to promote the use of innovative technologies in the DOE-EM complex. As of April 1998, top-level oversight was provided by the EM Integration Executive Committee. The effectiveness of this upper-level management involvement is unknown to this committee.

CHAPTER SUMMARY

This chapter's brief history of EM underscores the impact on OST of the range of broad-based priorities, the emerging importance of key management personnel in DOE in setting national priorities such as risk and mortgage reduction, and the significant shifts of direction for cleanup efforts across the DOE complex (i.e., from a long-term to a shorter-term focus on user needs). The evolution of OST program units reflects a growing awareness of barriers to the deployment of new technologies. Part of the rationale for the reorganization of program units over the years was a recognition of the deployment issue as a performance measure and as an EM-wide internal problem. Hence, OST has attempted to facilitate deployment of new technology with more targeted program structures addressing internal barriers and incentives to change.

A fundamental question remains—whether the DOE-EM climate still lacks the proper incentives to encourage the use of innovative techniques. Previous NRC reports (NRC, 1995a) and other sources (U.S. Congress, 1997; GAO, 1994; NETAC, 1995) addressing this point have discussed characteristics of DOE-EM site cleanup activities that may discourage such use. These include a culture that avoids taking chances, an unwillingness by site managers to change an already-established baseline remediation approach, a perception of job loss by the work force once the environmental waste cleanup problem is declared solved, and an unwillingness to jeopardize established stakeholder agreements. Such characteristics imply that the barriers to deployment are very complex and are at least in part external to EM-50.

The limits of OST's authority and interactions with other program offices within DOE-EM define an institutional environment whose complexity provides an important context for OST decision making. These issues are treated in Chapter 4, following an examination of practices useful to other organizations in industrial, nonprofit, and/or environmental technology development settings.

3

RD&D Decision-Making Practices in Other Organizations

This chapter provides a discussion of good decision-making practices that are drawn from literature surveys and benchmarking studies of major U.S. companies and from committee member interviews with representatives of organizations with significant environmental remediation programs. This chapter covers a range of applicable good practices and methods used in industrial RD&D and in two nonprofit research organizations whose scope and operating environments are analogous in some ways to those of DOE-EM. Practices in profit-making organizations at some stages are generally similar but are subject to greater differences in criteria, operating environments, and measures of success. The wide range of inputs from many different organizational contexts provides a good basis for recommendations for refinements in the methods and practices used in DOE. Those practices that have particular relevance to OST are highlighted.

BENEFITS OF A HIGH-QUALITY DECISION PROCESS

Decisions are the product of management work. They are arrived at through a process, although that process is not always explicit. Decisions have customers and should conform to customer requirements.

Research and application experience has shown that there are many benefits from using a high-quality decision process (Matheson and Matheson, 1998; Matheson and Menke, 1994; Menke, 1994). Process benefits include the following:

- depolarizing high-conflict situations potentially involving strong differences of opinion,
- ensuring comprehensive consideration of relevant factors,
- ensuring consideration of a wide range of alternatives,
- leveling the playing field across multiple projects and decisions,
- maintaining a consistent approach as people change positions,
- developing consensus and building commitment to action,
- facilitating explanation of the decision to internal and external parties, and
- providing documentation for retrospective analyses and insight on post-project audits.

These benefits come not only from the nominal decision process itself, but also from the skill and quality with which it is carried out.

To expand upon the last item, one important benefit of a well-structured process is to facilitate learning from past experience, leading to continuous improvement in the decision process. Decision research has shown that failure to learn from experience and to apply the lessons learned is one of the most pervasive and harmful problems in decision making (Russo and Schoemaker, 1989).

A decision process incorporating some of the best practices should lead to better results. Some evidence to support this conclusion is presented next.

LESSONS FROM INDUSTRIAL RD&D ORGANIZATIONS

Because of RD&D's importance to long-term success and sustainable competitive advantage, many industrial organizations have devoted a significant effort to developing and improving their RD&D decision-making process. This allows one to learn from a group of organizations that allocate substantial resources to RD&D and that can be judged to have achieved significant and valuable results from this expenditure. Benchmarking high-performance organizations to determine their "best practices" is a widely used approach for such learning.

An extensive research project on how outstanding RD&D organizations actually make decisions defined 45 best practices (Matheson et al., 1994), which are presented in Appendix F and summarized below. Although more than 300 RD&D organizations participated in this project, a group of 79, selected on the basis of peer group nominations, was designated as the "best company database" and used to quantify the importance and implementation of these practices. The use of these best practices is clearly associated with higher performance as indicated by the number, revenue, and profitability of new products (Menke, 1997a).

As a further result of work by Matheson et al. (1994), 10 of these 45 best practices were shown to stand out as especially important by having a potential contribution to decision quality with a mean score higher than 6.0 (on a scale of 1 to 7, with 7 as the highest possible score). These 10 practices have been identified as essential for RD&D strategic excellence in industrial organizations (Menke, 1997a), and are listed in Appendix F.

Further analysis has identified 10 other practices for gaining competitive advantage (Menke, 1997b), which are also listed in Appendix F. These are the 10 least well implemented from a set of 28 (out of the 45) practices judged to be very important in terms of performance impact or potential contribution. Because these 10 practices are important but not very well implemented (even by some of the most outstanding RD&D organizations in the world), organizations that want to gain a competitive advantage or improve their operational efficiency[1] can very likely do so by implementing these additional practices.

The committee's conclusion, based on the information in Appendixes B-E, is that many of the same practices should be relevant to making good RD&D decisions in a public-sector RD&D organization, such as DOE OST. There was general agreement among committee members as to which practices should be most relevant. These are discussed in the next section.

INSIGHTS FROM THE BENCHMARKING STUDY

Using the descriptions of best practices shown in Appendix F, committee members selected what they believed where the most important practices for OST to achieve excellent technology RD&D decisions. There was very strong agreement among committee members' selections, and the practices that the committee members felt were most important were, with only a few justifiable exceptions, the same as the 10 essential practices mentioned above. The most important of these was the following:

- Hire the best people possible and maintain expertise.

---

[1]Since DOE-EM is a government organization and therefore not *a priori* in a competitive environment, an appropriate analogy to gaining competitive advantage might be improving efficiencies of operation. On a technical level, any RD&D result must "compete" with prior and alternate ways of doing something.

Committee members thought that this applied to OST as well. However, these general statements beg translation to be appropriate in the OST application. Since OST is essentially a federal program of DOE program managers, who award funds to contractors to perform technology development projects, a suitable translation of this thought for OST might be, "Select the most competent contractors." The committee prefers to leave these translations to OST management, who are well positioned to appreciate how best to implement these general guiding ideas.

Committee members thought that dealing with end customer needs was also critical for OST, as represented by the following:

- Understand, focus on, and monitor changes in customer needs and requirements.

This statement combines the ideas of 3 of the top 10 essential practices: (1) focus on end customer needs; (2) determine, measure, and understand end customer needs; and (3) refine projects with regular customer feedback.

The needs-focused approach helps ensure that the technologies are relevant to customer requirements. Successful companies make every effort to determine, understand, and measure their customer's needs. Their RD&D organizations solicit frequent input. Many techniques can be used to gain customer feedback and to inspire customer support and loyalty. The key is a *structured* interaction process that returns knowledge, insight, and information. That is, the structure advocated here is that of having a well-defined (e.g., recorded) process by which (1) needs are assessed, (2) new projects are evaluated, and (3) funds are allocated. For example, criteria used to evaluate proposed projects should be adequately descriptive so that their meanings are clear to all interested parties. This structure is not intended to constrain or limit inputs from certain sources; indeed, the decision-making process should allow for all appropriate inputs. Rather, structure enables the process to be widely understood and critically examined for improvement over time.

Two other essential decision practices that committee members agreed were very important for the DOE-EM environment are the following:

- Agree on clear and measurable goals.
- Use a formal (i.e., common, consistent, structured, and rational) technology development decision-making process.

To have the greatest likelihood of success, several management levels should share strategic objectives. Ideally these goals are directly related to the prioritized needs of the customers of RD&D, needs that are related to the goals of the user organizations (in this case EM-30 and EM-40). The most successful companies also use a formal RD&D decision-making process with phases, checkpoints, and milestones to frame decisions and track their implementation. The process includes features and documentation that make continuous improvement possible through learning and iterative refinement.

There are 17 other decision practices out of the full 45 that the committee thought could be particularly important to the goal of improving OST's decision-making process. They fall into several important categories:[2]

- Think strategically (i.e., long range and high impact), which combines the following ideas represented by the best practices:

    1. Frame RD&D decisions strategically.

---

[2]These statements may read like general platitudes but, again, are offered for illustrative purposes as descriptive of methods that work well in other institutions; OST management would have to evaluate how far these concepts can be applied within the OST program.

2. Understand the drivers of change.
3. Insist on alternatives
4. Coordinate long-range business (for DOE-EM, the analogue to "business" would be the "site user") and RD&D plans.
5. Balance innovations and incremental RD&D.
6. Hedge against technical uncertainty.

- Measure and evaluate to guide resource allocation, which includes the following:

1. Ensure credible, consistent inputs.
2. Quantify decision inputs.
3. Measure the contribution to strategic objectives.
4. Evaluate the RD&D portfolio.
5. Manage the pipeline (and communicate how this is done).
6. Evaluate projects quantitatively.

- Communicate across organizational boundaries, which includes the following ideas of best practices:

1. Use cross-functional teams.
2. Coordinate development with commercialization (i.e., deployment).
3. Maintain intimate contact with internal customers.

- Continually improve the RD&D management process, which includes the following:

1. Learn from post-project audits.
2. Measure RD&D effectiveness.

"Think strategically" encompasses six of these decision quality best practices to stress the importance of a long-range strategic viewpoint in developing an effective RD&D program. "Measure and evaluate to guide resource allocation" encompasses six more. "Communicate across boundaries" refers to more open, cross-functional communication, which has been a major factor in improving industrial RD&D productivity over the past decade. The last category, "continually improve your RD&D management process," deals with improvement that comes from evaluations and audits of ongoing and past work. The practice of learning from post-project audits is advisable even for simple, straightforward decisions, and can be facilitated by keeping consistent logs or narrative records of the main criteria, factors, and measurements that were considered and evaluated. The recording formality is not just useful for supporting the decision-making process *per se*. Its other functions are to provide a coherent record and database for consistent administration when personnel or organizational assignments of responsibility change in the course of the project. Most important, the records also provide the basis for deriving "lessons learned." Decisions are rarely "optimum" when viewed in hindsight[3]. The log of the decision and implementation process can be used to reinforce good choices and practices in other projects, to help others foresee and avoid repeating the inadvertent pitfalls that are encountered, and to benefit from the preventive and remedial measures developed.

---

[3]Indeed, optimization is difficult in part because it is rarely, if ever, possible to know all the alternatives and their outcomes. The decision process then strives for a "good enough" result among likely candidates of relevant alternatives (March, 1994; 1999; Simon, 1997).

Decisions always have to be revisited after changes in goals or missions are promulgated, whether by the industry, agency directorate, or Congress. This requires re-evaluating the alternatives to be considered, the information available, and the applicable criteria. Whenever there is a change (or even a more crisp definition) of goals, criteria for success, and measurement methods, many prior decisions become subject to change. For decisions in such less-than-simple conditions, the use of a formal decision process (including record-keeping practices) becomes prudent, even for instances in which it appears to some participants that one or two well-understood factors dominate the decision (typically these are budget cuts or line-item directives). At stake is not merely failure to achieve optimum results. An obvious hazard in decision making is the case in which an administrative decision maker believes that one or two "obvious" decision factors or goals are so dominant that the decision is automatic and requires no further thought. This can lead to prolonged pursuit of poor or ineffectual policies. There are many historical examples of large, persistent, and costly blunders in government, industry, and the military. [4]

A favorable outcome does not necessarily reflect an optimum or even a good decision practice, nor does a less favorable outcome necessarily reflect poor decision practice. An important measure of a successful decision is that its most likely outcome(s) are consistent with the stated intentions (goals) of the project and of the organization charged with responsibility.

The outcome of a decision may only partly fulfill the intended goals. Limited or no success in producing the expected outcome of a decision is not necessarily a sign of a flawed decision process. A basic measure of good decision-making methods and practices is that the decision remains "robust" in hindsight, regardless of outcome. For this purpose, the term robust means that the decision considered all possible outcomes and their probabilities of occurrence and that a better decision was not available *with the information available or obtainable at the time*. This recognizes the commonsense observation that almost any decision may be found to be suboptimal as further information is acquired or develops.

The valuation of possible outcomes should involve the group or community that is most directly affected by the decision. In a business context, the most likely outcomes are evaluated in terms of business values—for example, effects on market share, profitability, development of a viable new product or service, or better image as seen by clients. In the public sector, evaluation commonly also includes criteria such as equity, meeting stakeholder perceived needs, cost-benefit ratio, and both short-term and long-term impacts.

In selecting and guiding RD&D, many decisions occur over extended periods. For example, a selected technology development project is subject to continuing measurement and re-evaluation for both performance and consistency with the ultimate goals the project is intended to address, as it moves through states of development. In the evaluation of a project over time, as it matures from an initial proof-of-concept experiment to a large-scale demonstration of a properly designed and engineered prototype, better performance of the method under development is only part of the evaluative challenge. Well-managed RD&D is subject to continuing refinement of the approach being taken; that is, the organization's view of the optimal technology solution to a given need or problem can change during the development of a particular technology designed to address the need. The original decisions—on goals, budgets, measures of performance, relative likelihood of success and of application, and the priorities of various needs—are open to change as new information becomes available. The decision to continue, modify, or stop a given technology development project may arise from a reassessment of the need for it irrespective of the project's performance. Overall decision quality is achieved by periodic examination of these issues, resulting in a succession of decisions made over time on the scope and duration of any technology development project.

---

[4] *The March of Folly* by Barbara Tuchman recounts examples of large organizations and governments that persisted in decisions or policies that were manifestly and grossly to their own disadvantage, given any reasonable view of plausible intended goals (Tuchman, 1992; see also Scott, 1998).

INSIGHTS FROM INDUSTRIAL ORGANIZATIONS WITH ENVIRONMENTAL PROGRAMS

The committee was fortunate in being able to interview key environmental management executives from several private-sector organizations that have large environmental programs including significant remediation needs. Outside institutions were selected based on several criteria, including the following:

1. complexities in institutional structure that offer some analogue to OST within DOE-EM;
2. a larger community of interested parties that could be analogous to the institutions and individuals interested in DOE-EM site activities (e.g., the technology users at DOE-EM sites, the sites' surrounding communities, regulatory authorities, and the U.S. Congress);
3. a large RD&D program in technological areas similar to those of the OST program; and
4. a large environmental program with at least the potential for significant remediation efforts at multiple sites.

The committee could examine in depth only a limited number of institutions; by selecting the ones listed in this chapter, other reasonable candidates are not represented, and their omission from the few specifically mentioned here should not be construed pejoratively. Certainly other RD&D programs in federal agencies and private industries are valid models that offer insight into effective decision-making practices. The organizations selected and examined by the committee were Amoco, DuPont, Exxon, Mallinckrodt, Monsanto, the Electric Power Research Institute, and the Gas Research Institute. The focus of these interviews was not limited to technology RD&D selection, but covered the full scope of strategic decisions for the environmental programs of these organizations. This was necessary since several of these companies are themselves developing new or improved environmental cleanup methods only when necessary; that is, when commercially available methods will not suffice. Although some of the environmental challenges and consequences of the bureaucratic environment faced by DOE-EM are unique, DOE can still learn from the environmental decision-making and technology development practices of these leading industrial organizations.

Practices for Environmental Technology Development Decision Making and Management

The practices relevant to DOE-EM that the committee learned from more than one of these organizations include the following:

- Link environmental decisions to the business planning process as much as possible.
- Use qualitative and quantitative technical risk assessment approaches (with varying degrees of formality).
- Do a sensible prioritization of needs, and tackle the most costly and complex jobs first (good for "mortgage reduction"). For application to OST, this would mean putting resources in places of maximum benefit or return that is, in areas where technological innovation would have the greatest impact on cleanup costs or risks.
- Do not always wait to be forced to address environmental issues; it can be much cheaper to solve them voluntarily.
- Try to standardize the technologies used for recurring problems (do not "reinvent the wheel"). For application to OST, this would mean building and using a catalogue of "best technological approaches" that would provide a technological baseline of practices.
- Try to develop a proactive relationship with regulators, including seeking pre-approval of technologies for specific needs.

- Share insights learned and technology development costs with other companies or organizations with similar problems to the degree possible, through industry associations such as the American Petroleum Institute and the Petroleum Environmental Research Forum.

The first three environmental decision-making practices relate directly to several of the best practices for excellent RD&D decision making discussed above, even though they did not specifically refer to technology development in these companies. The first is almost identical to the essential practice "coordinate long-range business and RD&D plans," whereas the second and third provide some additional specificity such as the use of risk assessment approaches and prioritization based on the magnitude of the problem to be solved. The fourth, however, is a new insight. If problems are dealt with before "solutions" are externally imposed by regulators and stakeholders, the range of options available is usually much broader and much more cost-effective. The converse of this can be seen at several DOE sites where the most cost-effective technical solutions are precluded by legal agreements, such as the triparty agreements operative at Hanford.[5] A proactive approach could be valuable for DOE in going forward to the extent that flexibility still exists in some of its problem areas.

INSIGHTS FROM VISITS TO DUPONT, EPRI, AND GRI

The committee sought input from large institutions that in some ways could be compared to OST within DOE-EM. These institutions had as common features a centralized technology development program (the counterpart of OST) and separate, quasi-independent "business units" or "member utilities" (the counterparts of other EM offices, such as EM-30, EM-40, and EM-60). Another common feature of these analogous institutions was a context of external regulation in an industry faced with some environmental cleanup issues that served as a basis for technology development activity. These features were criteria in selecting institutions for study. The purpose of gathering information from these organizations was to probe their decision making to discover effective practices that might lend themselves to adoption by OST.

Three institutions—DuPont, EPRI, and GRI—were chosen for in-depth scrutiny, based on the above criteria. During field trips to these organizations, committee members interviewed upper-level managers from the technology development programs. The information learned from these interviews and from published company literature is described in detail in Appendixes G, H, and I. Salient points that the committee extracted for emphasis and application to OST are summarized below.

Institutional Structure

There is nothing wrong *per se* with individual business units' or member utilities' (hereafter referred to as companies; their EM counterparts are the other EM offices, such as EM-30, EM-40, and EM-60) having their own internal RD&D operations. In each institution, a healthy, long-term relationship was established by which certain RD&D projects (those of a short-term nature or tied to specific company processes) were channeled to each company's internal RD&D group, while other RD&D projects (those representing common and long-term needs, with universal application to many companies) were directed to the central RD&D program. The internal RD&D groups of the separate companies or divisions tended to have a narrow focus on unique, local problems and on RD&D projects with a short time horizon. These projects are better characterized as development work rather than as RD&D. Since the central RD&D facility does not fund

---

[5]It is widely recognized that some of the legal agreements involve high marginal cost and low risk reduction benefits for specified actions. The use of RD&D to identify alternative technical approaches has merit in these situations, as discussed further in Chapter 5.

every proposal directed to it, the internal RD&D operations of the individual companies are necessary to ensure funding of proposed work that is truly critical to only one or two companies, that is, that they can meet their own immediate needs.

Effective Top-Level Strategic Goals to Define the Suite of Technology Projects

Another feature common to DuPont, EPRI, and GRI is that top-level, strategic goals are cast in measurable, quantitative terms and are effective drivers to define the suite of technology development projects proposed for funding. These goals were written as achievements for each separate business unit to strive to accomplish in the time frame of a few years. For example, DuPont set a goal in 1993 to reduce toxic air emissions from its domestic facilities by 60 percent over three years from the 1993 benchmark level. Similarly, one of GRI's program units developed a goal to "develop and transfer technologies that lower the costs of gas-well drilling by 5 percent by year 2000," (GRI, 1997b), commensurate with its more top-level objective to keep natural gas competitive with other energy sources in consumer demand. Such goals then defined the types of RD&D proposals that were generated by researchers and technology developers and entertained for review by program managers. RD&D goals and performance specifications are selected after careful negotiations between the "marketers," who survey and determine the needs of clients, and the "developers," who intensively estimate the results available for a given time and cost. This provides a guarantee to potential sponsors of the RD&D efforts that the results will be provided at a specified cost and schedule.

Measure and Evaluate to Guide Resource Allocation via Return-on-Investment Evaluations

The centralized technology development programs of DuPont, EPRI, and GRI seek to operate in an objective, matter-of-fact environment in which the selection of projects to fund is a business decision. Hence, the scoring of proposals by some appropriate ROI measure (e.g., the GRI Project Appraisal Methodology scoring described in Appendix H or the DuPont method of converting consequences to dollars as a way to provide a relative ranking) is crucial to informing the decision of which funding candidates to select. As reported to committee members, these three organizations perform an ROI-type calculation on each technology development proposal to compare all the new proposals in any given year prior to the annual funding commitments.

Customer Buy-In

At GRI and DuPont, the central technology development program was funded by the parent organization with funds separate from those of the business units or member utilities. In these institutions, the parent organization's funds came from revenue or dues from each business unit or member utility. By contrast, the EPRI member utilities that wanted access to the products of specific technology development activities provided the funding for those activities. Both of these approaches achieved member buy-in, because in both cases the business unit or member utility proposed ways to spend technology development funds that, directly or indirectly, came from its contributions.

PROCEDURES AND METHODOLOGIES TO SUPPORT GOOD DECISION MAKING

Sources such as Matheson et al. (1994) and the DuPont, EPRI, and GRI interviews indicate the importance of measurement and the use of structured evaluation methods to support RD&D project prioritization and decision making. The committee's selection of best practices most relevant to the DOE-EM situation included six directly concerned with measurement and quantitative evaluation. In interviews with industrial organizations, the importance of qualitative and quantitative technological risk assessment techniques was stressed. Finally, the detailed site visits to GRI, EPRI, and DuPont revealed further evidence of the use and value of quantifiable, probabilistic cost-benefit evaluations to help prioritize projects and allocate resources for maximum benefit. GRI in particular demonstrated that this could be done in a multicriteria setting; DuPont's experience showed that such methods can help achieve the maximum environmental cleanup per dollar expended; and the experience of many excellent industrial organizations indicates that such quantitative methods can have a powerful impact on improving the return on the RD&D expenditure.

Utility and Limitations of Quantitative Evaluations

The committee believes that structured methods and evaluation criteria can play a very important role in the DOE-EM technology development decision process, subject to several very important caveats. First, the quantifiable methods and results should not be viewed as making the decisions, but rather as providing very important input to the decision-making process that must be discussed and reviewed in light of all relevant factors. Second, a poorly done quantitative process is worse than no quantification at all, and often, effort and money can be spent generating poor or even misleading numerical results. Third, quantifiable results must be judged by recognizing possible weaknesses in the methodology used, uncertainty in the numerical inputs, and important decision factors that either were considered unquantifiable or were not quantified.

There is often justifiable resistance to quantification since quantifying poor information may lead to misplaced confidence in the results. For decisions involving large allocations of resources, modern decision methods (e.g., Bayesian analysis) should be used, since this also quantifies the uncertainties and displays their influence on the overall decision or outcome. If uncertainty is unacceptably high, it forces deferral of the decision and additional work to improve the quality of the information inputs.

The large body of methodology and experience encompassed by decision and risk analysis, probabilistic risk analysis, and cost-benefit analysis provides a wealth of procedures, methodologies and tools to implement a sound, carefully structured approach. These approaches have many common features, such as (1) focusing on finding the most productive ways to allocate scarce resources; (2) measuring benefits explicitly; in quantifiable terms whenever possible; and (3) dealing explicitly with uncertainty and technological risk using probabilities. A project evaluation system embodying these principles and applied consistently across the potential projects to be funded would help greatly to improve decision making at the project funding level, as well as provide insight into portfolio evaluations, budget allocations among major program areas, and the effectiveness of the total OST budget. The published experience of Eastman Chemical Company, which has applied quantitative evaluation to guide its entire RD&D quality program, demonstrates a doubling in the expected value of the RD&D results accomplished over just a few years (Holmes, 1994).

To date, OST has attempted to (1) address productive ways to allocate resources by emphasizing user input to the decision-making process and (2) measure benefits by counting technologies implemented and estimating cost reductions from using alternative-to-baseline technologies. There has been no apparent attempt to deal with uncertainty or technological risk using probabilities. These OST applications, including a way to treat uncertainties using probabilities, are discussed in more detail in Chapter 5, after Chapter 4 outlines the decision process called for and how OST has implemented the requisite process steps.

CHAPTER SUMMARY

Although a plan for improving decision making must be tailored to the needs of a specific organization, the results (Matheson et al., 1994) from benchmarking highly regarded industrial and consortium RD&D organizations clearly demonstrate some basic points that the committee thinks are relevant for government as well as industrial organizations.

There are decision-making practices, well established in leading industrial organizations as well as in EPRI and GRI, that can contribute to decision quality in an RD&D environment. High-performance organizations are more likely to implement most of these practices better than organizations with lower performance, especially practices related to making quality decisions (see Appendix F for a list of these practices). Since decision processes exist in the context of an organizational structure, some of the practices apply to RD&D management, whereas others apply to the decision process itself.

Drawing on published practices, the major ones that the committee recommends DOE-OST focus on are the following:

- Understand, focus on, and monitor changes in customer needs and requirements.
- Agree on clear and measurable goals.
- Use a formal (i.e., common, consistent, structured, and rational) technology development decision-making process and apply it uniformly.
- Think strategically (i.e., long-term and high impact).
- Measure and evaluate to guide resource allocation.
- Communicate across organizational boundaries (i.e., with technology users).
- Continually improve the research and development (R&D) management process.
- Hire the best people possible and maintain expertise.

Each of these recommended practices encompasses one or more of the best decision-making practices listed in Appendix F. However, it is also important to note that RD&D organizations use a strategic management process that embodies many of these practices in a mutually reinforcing manner (Lander et al., 1995); therefore, OST should not expect that simply selecting one or two practices for implementation will be adequate to achieve excellence. These concepts have to be translated into appropriate statements for OST, a translation that the committee prefers to leave to OST management.

The results from industry environmental decision making and the planning and budgeting processes of GRI and EPRI demonstrate the use of such good practices in settings analogous in some ways to OST. Therefore, the committee thinks that these best practices are relevant to DOE's technology development decisions. The committee has identified the practices thought to be most relevant to DOE and finds that there is substantial room for improvement by OST, as indeed there is in most of the industrial organizations benchmarked by Matheson et al. (1994).

The descriptors of the best practices used here are fashioned to capture one or more related ideas taken from already published work and are offered here to describe the attributes of a successful RD&D management environment. These descriptors are written in general language, but it is not the committee's purpose to translate the way each should be applied to OST, to evaluate OST against these statements, or to define all appropriate contexts in judging how well they apply to OST. Rather, the intent is to identify practices that OST management can use in informal internal evaluation (and improvement) of the prioritization and decision-making activity within the OST program. In the next chapter, the present status and realities of the DOE-EM situation and its limitations are examined in light of these best decision-making practices.

4

Decision Making in Research, Development, and Demonstration for the DOE-EM Program

The key decisions for a DOE-EM technology development program are nominally those to identify and prioritize technology needs and to solicit and fund projects in support of these needs, with the aim of helping to achieve EM cleanup goals. The way this is done within a federal program such as DOE-EM is governed by the framework (e.g., the federal budget planning process) and constraints (e.g., no direct control over the use of technology) of the institutional environment in which a central RD&D program office such as OST operates.

These considerations are treated in this chapter with a discussion of the major decisions that must be made at a functional level, followed by a description of OST's institutional context. This context provides constraints, both internal and external to DOE-EM, that have some influence on whatever procedures are developed to perform the requisite functions. These procedures are a sequence of process steps that are institutionalized and conducted by the RD&D program office (i.e., OST) to make decisions. A decision-making process is presented below in a general form and then discussed in the context of current OST program structures. This chapter describes how the various process steps were conducted within the current DOE institutional structure; evaluative comments are provided in Chapter 5.

MAJOR DECISIONS AT A FUNCTIONAL LEVEL

Table 4.1 shows the major technology development decision-making activities, which are conducted at various hierarchical levels in the federal organization. The first and last organizational levels conducted are outside of OST. The decisions made at those levels represent important influences outside the direct control of the RD&D organization.

RD&D PROGRAM ENVIRONMENT

The institutional environment, jurisdictional boundaries, and factors that impact the RD&D program within DOE-EM strongly influence the development of any decision process. OST decision making occurs in an environment with the following important features:

- Many players are involved, including Congress, the White House (via the OMB), DOE-EM technology users, stakeholders,[1] and OST.

---

[1]As used here, a "stakeholder" is anyone outside the DOE organization who is interested in or affected by the activities of the EM program. Important stakeholder groups whose input is solicited by the EM program are local

TABLE 4.1 Decision-Making Activities for DOE-EM RD&D

Organizational Level	Decision-Making Activity
Within the Office of Management and Budget (OMB) and Congress	Establishing budgets for DOE-EM RD&D
At DOE-EM headquarters, within the RD&D program	Establishing strategic objectives Proposing budget, and corresponding responsibilities and goals, among separate technical areas (e.g. Focus Areas and Crosscutting Programs)
Within RD&D program units	Determining what technology development work needs to be done (this requires interactions with the technology users and generation of a prioritized list of development needs across the EM sites, to be used as a basis for drafting requests for proposals) Evaluating proposals to select the technologies to be funded Evaluating and monitoring progress of funded projects
Within the site remediation organizations (i.e., the other DOE-EM offices that have remediation and waste management responsibilities at DOE sites)	Developing a baseline functional flowsheet[a], and alternatives for each waste stream or remediation problem Determining the technology needs required to implement the baseline functional flowsheet (and alternatives to it), and forwarding these needs to the RD&D organization Determining whether the RD&D organization's technologies (i.e., those that have been or will be developed and demonstrated) will meet all the functional flowsheet requirements Deciding whether to use these RD&D technology products (i.e., their deployment)

[a]The baseline functional flowsheet is the presently preferred sequence of process steps that comprise the waste treatment sequence from the initial waste configuration to the final waste end state. It defines the technology needs in a general sense. A baseline functional flowsheet represents major commitments by the site operators to a technology path in which a great deal of planning and engineering commitment has been made and, very often, large amounts of money have been spent. In addition, the site operators have numerous nontechnical constraints, both regulatory and political, that discourage technology changes. Nonetheless, significant improvements to baseline functional flowsheets are possible.

- The federal budget process requires monitoring the current year budget and planning at least the one-year-out and two-year-out budgets. Therefore, three budgets have to be managed at any one time, in varying degrees of detail and certainty.

- OST is neither subordinate nor superior to other EM offices, which have cleanup responsibilities for the sites. In practice, this means that OST has no way to force the use of technologies it develops. However, these technologies must be used at DOE sites for OST to justify spending funds to develop them. Remediation "functional flowsheet" decisions are made by DOE sites funded by these other EM offices in response to regulatory and other drivers.

- Top-level EM goals[2] and planning exercises can influence the activities of OST and other EM offices. The user-supplied needs that OST responds to are frequently driven by EM goals (and other influences) that, when changed, result in changed technology needs.

---

advisory boards, comprised of citizens living near major DOE-EM sites; Native American tribal nations whose land is affected by DOE-EM activities; and groups such as the Community Leaders Network and the Western Governors Association. The taxpayer at large, as represented by Congress, also has an interest in reducing the costs associated with remediation activities.

[2]Headquarters priorities, such as risk and mortgage reduction, are examples of how top levels of management can dramatically change the timing of technology needs through changing goals. The use of risk reduction as a priority-

- Congress makes a decision annually to provide OST's funding based on broad perceptions of OST's benefit to the nation. Thus, OST has Congress as a "customer" as well as the technology users. On occasion, Congress provides directed funding to particular projects or organizations (U.S. Congress, 1992; 1994; 1995a-b).

MODEL PROCESS STEPS FOR DECISION MAKING IN A CENTRALIZED FEDERAL RD&D PROGRAM

The model in Figure 4.1 describes the managerial functions of an RD&D organization in the federal government. This figure is a generalization of the processes described in Appendixes B-E, constructed from site visits, documentation, and committee members' collective experience. It shows two important features of OST's current decision-making process: external constraints and functional steps, discussed separately below.

External Constraints

Figure 4.1 diagrams the way OST interacts with the chief external organizations that provide important jurisdictional boundaries, namely, other DOE-EM offices and funding authorities, as depicted on the left-hand side and in the top section, respectively. The processes represented within these sections are outside OST's direct control but have impact on its decisions.

Functional Steps

The boxes on the bottom right-hand side represent functions that must be performed in a decision-making approach of the kind OST has adopted, in which the RD&D program is based on site technology needs. The specific functions of this process are depicted as boxes representing the following steps:

- site technology needs are acquired from the user community (Box 6);
- these needs are prioritized (Box 7);
- a plan is developed for technical activities based on these prioritized needs and the available budget (Box 8);
- Projects are solicited , and responses are evaluated to select those to be funded (Box 9);
- the principal investigator (PI) of each funded project conducts the work (Box 10); and

---

setting measure governing EM work was championed during 1994-1996, corresponding with Thomas Grumbly's tenure as the DOE-EM Assistant Secretary. Beginning in 1996 with Alvin Alm's tenure as the DOE-EM Assistant Secretary, mortgage reduction—buying-down near-term costs by putting facilities in low-cost maintenance conditions—was given highest priority. Either idea is readily incorporated in the selection of technology development projects and may be emphasized by increasing its relative weight compared to all other criteria against which projects are scored. It is clear that both risk reduction and mortgage reduction are worthwhile and necessary goals with comparable importance. However, the goals may be in conflict when compliance agreements are excessively rigid, for example, when remediation is required that results in negligibly small reductions in risk. This is a critical area of policy to be resolved by DOE and Congress.

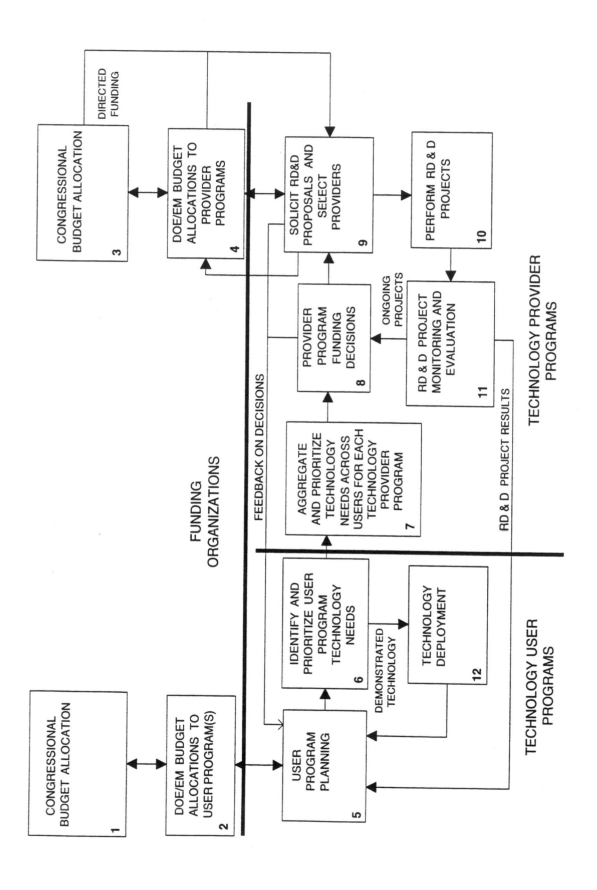

FIGURE 4.1 Idealized DOE-EM decision-making process for RD&D projects.

- OST monitors and evaluates (Box 11) progress until funding ceases.

A more detailed account of these functional steps, including a description of how they are currently done, is provided in the rest of this chapter.

Allocation of the Budget Among User and Provider Program Units

All DOE-EM program offices develop plans as a part of the federal budget cycle and receive their budgets from congressional appropriations. At the headquarters level, DOE upper management proposes the allocation of program funds among several subordinate program units. This allocation is then approved by Congress, often after modification. Occasionally, as in the case of the Environmental Management Science Program, Congress directs the appropriation of specified budgets to selected programs or projects.

The steps in this part of the model decision-making process are depicted in Boxes 1-4 of Figure 4.1. Boxes 1 and 3 show congressional decisions on budget allocations to two different types of EM program offices. Box 1 shows allocations directly to a technology user program, so-called because its RD&D-related role is that of the user of any RD&D results. Examples of user programs are DOE-EM offices such as EM-30 and EM-40 that bear responsibilities for site cleanup and waste management activities. Box 3 shows allocations to another type of program office, a technology provider program such as OST, which funds technology development activities and in so doing must interact with and be supportive of the user programs if it is to be effective and demonstrably beneficial.

Boxes 2 and 4 show the role of DOE upper management in deciding what program funding for separate subordinate program units to propose in the budget submitted to Congress. For a technology user such as EM-30 or EM-40, these subordinate program units would be DOE field offices serving EM sites; for a technology provider such as EM-50, these program units would be Focus Areas or Crosscutting Programs.

Planning Activity Involving Technology Decisions

The EM offices plan during any one fiscal year for program work that may extend into the future over many fiscal years. Based on its own planning activities (Box 5 of Figure 4.1), the user identifies technology needs (Box 6) to support its site remediation functional flowsheet plans. These needs are sent to the provider. Because a typical technology development project takes at least two to three years, the provider must receive these needs sufficiently far in advance to allow time for development work. In practice, this means that the user needs for technologies in the future affect provider program development for more immediate budget cycles.

Process Steps Internal to the Provider Program

After receiving technology needs as inputs from a user program and a budget from upper-level management, the rest of the provider process steps are managed by the provider program. The provider manages the processes to

- collect and prioritize user technology needs[3] (Box 7);
- decide which technology needs to develop into funding solicitations (Box 8), which includes as substeps the identification of technical solutions to these needs statements and recognition of those technologies that may already be available outside DOE so that "make-or-buy" determinations can be performed (it is at this step that care must be taken to avoid unnecessarily duplicative technology development projects);
- prepare and tender solicitations for technology development proposals and evaluate the responses to select the ones to fund (Box 9);
- initiate technology development projects by providing funds to the PI of each proposal selected (Box 10); and
- monitor these technology development projects (Box 11) to evaluate their progress over time.

Deployment

A very important aspect of the user-provider program interactions (Figure 4.1) is that deployment (Box 12) is controlled by the user. For those projects that are not terminated during the RD&D process for reasons such as technical difficulties that prove to be intractable, budget shortfalls, or changing needs, the end point is deployment of the technology to address DOE-EM's problems. This activity normally occurs when a DOE-EM problem owner determines that

- sufficient technical knowledge exists to design, build, and operate the new technology;
- an adequate budget is available for deployment; and
- the problem owner's need for such new technology has persisted.

The problem owner (i.e., a technology user) can determine that the technology is adequate based on

- information received from documentation provided by the technology provider and from the user program office's participation (e.g., by cofunding the final stage of technology development) or
- the technology's demonstration in a field environment or on real or simulated waste, on a large-enough scale to foster confidence in its real-world application.

Necessary Interactions and Feedback

The arrow from Box 11 to Box 5 of Figure 4.1 represents, for each funded technology development project, the ideal outcome: demonstration of a technology that the user community then adopts.

The arrow from Box 9 to Box 4 shows the role of each provider program unit (e.g., each Focus Area or Crosscutting Program) in presenting and defending, at the headquarters management level, information such as the program's funding choices (i.e., the suite of technology development projects that represent the program's current investment), the past deployment successes, and future strategy. Headquarters judgments on funding allocations (Box 4) are influenced in part by past and present performance and by future plans.

The arrow from Box 11 to Box 8 depicts a flow of information that is useful for making a decision on whether to continue or terminate an ongoing technology development project. Project termination prior to the completion of the work could occur for several reasons. One reason is poor performance by the PI

---

[3]Prioritizations of needs and of solutions are frequently inseparable, especially for ongoing projects or site-specific demonstrations.

(e.g., progress milestones are missed or funds are not spent in an appropriate or timely way). Research by the PI may show that the technology will not work as anticipated initially. Another reason for early project termination might be that the need it was designed to address may have become less important in another year's reprioritization.

Interaction between the provider program and the community of users of technology takes place on a continuing basis, as reflected in Figure 4.1, by the monitoring function, the results of which are communicated from the provider to the user as appropriate.

IMPLEMENTATION OF THE MODEL PROCESS

Good decision making (and the process steps of Figure 4.1) can occur within many possible program structures. OST's Focus Area-based structure was built on recommendations developed in 1993 (see Chapter 2).

The overall method used by OST for allocating RD&D funds can be described within the framework of Figure 4.1, primarily by Boxes 6-9. In this framework, the STCGs at each major DOE-EM site interact with representatives of the user programs to identify and prioritize technology needs at the site level (Box 6). The STCGs provide the Focus Areas with prioritized lists of technology needs. These inputs are used to develop the Focus Area's prioritized list of needs (Box 7), which will form the basis for solicitations for technology development project work (Box 8). Funding decisions on specific new projects proposed in response to solicitations are made through a review of these proposals (Box 9). The Focus Area's priority ranking of needs, with the determination of its budget (Box 4), determines the funding cutoff. More detail on the steps represented by each box of Figure 4.1 is presented in the following sections.

User Program Funding (Boxes 1 and 2)

Funding is provided to a user program office through the annual federal budget process. This process focuses on the two fiscal years following the year in which the budget is being formulated. That is, the budget deliberations for FY 1998 (October 1, 1997, to September 30, 1998) began with the preparation and submittal of a top-level budget to OMB and Congress during the spring of 1996, with a more detailed and refined budget prepared and submitted to OMB and to Congress during the spring of 1997.

For a long-term program such as DOE-EM, the budgets, plans, and forecasts are sometimes made decades into the future. Based on such future projections and past accomplishments, each user program office prepares budgets that are forwarded to successively higher levels of DOE administration up to the executive branch, and sent back, in a process with some iteration. Policy controls are applied by the OMB. The result is the President's two-year-out budget request to Congress in early January. Congress (and its staff) deliberates on this budget through formal hearings and informal direct discussions with upper-level DOE-EM management. Congress can then alter the budget package by increasing or decreasing funding, creating new programs, changing the goals of existing programs, or specifying in varying degrees of detail how funds are to be spent. The result is budget legislation (a law) that is sent to the President to sign, usually before the new fiscal year begins. The process repeats for the next fiscal year. The budget legislation provides the basis for DOE-EM program planning, which thus can be finalized only after the law is passed.

Provider Program Funding (Box 3)

One of the major forces that both drives and shapes DOE-EM decision making in its RD&D activities is the process by which OST's total annual funding is determined. Because DOE-EM is funded annually by appropriations from the U.S. Congress, the decision-making process must take place within the framework of the federal and DOE budget process. This process takes place over three fiscal years (FYs):

- *Program budgeting in FY-2.* During this year OST prepares a top-level budget to be executed two years hence in the program execution fiscal year. This budget is adjusted and approved within DOE to meet spending targets and then submitted to the OMB late in FY-2.
- *Program planning in FY-1.* The FY-2 submittal is often iterated at least once with OMB during FY-1, and there are frequently informal discussions between DOE and congressional staff members to define an acceptable budget. Then, on about February 1 of FY-1, the OST budget is submitted to the U.S. Congress as a part of the much larger President's budget. For OST, this submittal would include a specific figure for the total OST budget and the budget of major subprogram units (e.g., each Focus Area). In parallel with this, the Focus Areas are acquiring technology development needs from the STCGs, developing a prioritized list of responses to the needs, obtaining the necessary concurrence from user representatives, and issuing guidance to their operating elements. Usually, but not always, both the congressional budget process and Focus Area activities are completed before the beginning of the next fiscal year on October 1.
- *Program Execution in the current FY.* During this year the OST program outlined in FY-2 and in some detail in FY-1 is implemented. This includes formally sending funds to performers and monitoring ongoing work. Simultaneously, the activities described immediately above for FY-1 and FY-2 are proceeding for future years. Events during the year often cause either DOE or the U.S. Congress to adjust the budgets, usually downward via rescissions.

Therefore, at any one time, OST manages three budgets. One is the current fiscal year's budget, already allocated by Congress. The second is the upcoming fiscal year's budget, for which specific plans are prepared in the spring and forwarded to Congress. The budget of each OST program unit (e.g., Focus Areas and Crosscutting Programs) is represented at the level of specificity of "work packages," as defined and described later in this chapter. Negotiations with OMB and appropriate congressional committees result in final budget figures in the fall, ideally prior to the October 1 beginning of the fiscal year. In step with this process is the formulation of the two-year-out budget, represented at the level of specificity of the requested budget for each major OST program unit and used for planning purposes by these program units as budget targets.

For the provider program to be responsive to the technology needs of the user program, a technology need in the long-term plans of the user program would have to be addressed by a technology development project finished by the required time. Hence these technology development project funds are contained in more immediate fiscal year budgets of the provider program. As a consequence, a change in the user program plans for later years affects the plans of the provider program in more immediate fiscal years.

User or Site Program Planning (Box 5)

The user programs develop new plans or modify existing ones to accomplish cleanup objectives within their purview. These cleanup objectives are often required to meet compliance agreements. Plans are typically defined for an entire user program (e.g., remediate tanks containing high-level waste) and detailed at successively lower levels so that work for individuals or small groups eventually is defined.

Identification and Prioritization of Technology Needs (Box 6)

Planners of user projects must assess the capability of existing technology relative to what is required to achieve the specified objectives, such as specifications of the end state of the waste products (NRC, 1998a; 1999a) in terms of composition, form, location, and cost. If technology to meet these objectives is not available, this deficiency should define a technology need that user program personnel (such as site contractors) identify and forward to OST. OST collects these needs from site personnel via the STCGs (see Chapter 2 and Appendix B).

The STCGs are responsible for developing and prioritizing a list of site problems and technology needs. A technology need can exist for several reasons: a gap exists in the baseline plan for managing an environmental problem; an opportunity exists to replace a baseline technology with a better and less expensive one; there are several possible alternative technologies to the baseline, none of which has yet been fully proven or demonstrated; or there is no proven technical solution to a problem.

The STCGs provide the link between the problem owners at each site and the provider organization. At most sites, members of the STCGs are DOE staff, site contractors, and other stakeholders. Several STCGs have formed subgroups corresponding to the technical areas of the Focus Areas. Members of the subgroups are usually technical staff of DOE and contractors who are closely associated with the problem owners.

The STCGs describe technology needs in terms of the nature of the problem, relevant information regarding existing technologies and costs, operational and regulatory requirements, and technical performance. Each STCG has developed its own process and methodology for providing technology needs to the Focus Areas; specific descriptions and observations are made in Appendix B. Most STCG processes

- employ a user-oriented approach to technology needs identification,
- involve stakeholder input,
- use a structured and documented process for prioritization of needs based on STCG established criteria , and
- use a formal evaluation methodology.

The results of the STCG process are prioritized lists of technology needs that are delivered to the Focus Areas. The prioritized technology needs are described in a couple of paragraphs, and some qualitative reasoning behind the priority ranking or rating is provided, according to a reporting template agreed upon by all the STCGs.

Aggregation and Prioritization of Technology Needs for Each OST Program (Box 7)

Each provider program (e.g., a Focus Area or Crosscutting Program) selects a suite of technology development projects to fund. This is done by gathering and prioritizing technology needs from the STCGs and sites, usually with the approval of the potential technology users from relevant DOE-EM sites, a process described below in more detail.

Focus Areas use STCG needs statements and other resources to create a prioritized list of needs that is reflective of national, or EM-wide, priorities. Although the emphasis of the STCGs is by design on problems at a specific site, the emphasis of the Focus Areas is on a specific problem area across DOE-EM sites. The STCG needs statements are used to update the Focus Area's portfolio of needs and determine investment targets. If the Focus Area is aware of an available technology solution to a site problem, this information is transmitted to that site's STCG and the need is usually removed from further consideration.

Focus Areas typically aggregate similar individual needs statements to generate a smaller number of more general topical areas of need, which are then prioritized nationally. Each Focus Area has the flexibility to generate these broader categories of needs areas and prioritize them in its own way. For example, the Subsurface Contaminants Focus Area combined a larger number of site needs into approximately 18 work packages, which are groups of needs based on similar characteristics, written in general language (see Table C.1 and Figure C.2 for examples), and then prioritized into a smaller number of work packages. As another example, the Mixed Waste Focus Area determined national needs by using data from EM-30 Site Treatment Plans, which outlined mixed waste treatment capabilities. As a third example, the need to retrieve residual tank wastes at multiple sites has been combined into a multisite need of the Tanks Focus Area that is addressed by one robotics-related development effort.

Results of the needs evaluation and prioritization form the basis for building each Focus Area program's portfolio and for soliciting technology development proposals. Generally, high priority is assigned to the most important and urgent end user needs. The importance and urgency of end user needs in turn are derived from top-level EM and site priorities.

These provider programs solicit input from both the DOE-EM user community (i.e., other EM offices) and non-DOE-EM interested parties. The user community provides input to the OST STCGs in the form of statements of needs. Outreach to non-DOE-EM stakeholders is conducted by OST program units in different ways: Focus Area stakeholder outreach is through groups with a national focus, such as the Community Leaders Network; in contrast, the STCGs have a local focus only. OST's non-DOE-EM stakeholder outreach activities are generally to provide stakeholder input on valuations, with criteria and relative weights, to be applied to all technology development proposals (Appendixes B-C). It has not been OST's practice to challenge the "stakeholder agreements" struck with DOE-EM user programs or to use stakeholders as advocates of specific technologies.

Funding Decisions on Technology Needs and Projects (Boxes 4 and 8)

There are two types of budget allocation decisions in the overall framework described in Figure 4.1. The first allocates funds among the suite of provider program units (Box 4). The second allocates funds to specific technology needs and projects within each provider program (Box 8).

Allocations Among User Program Units (Box 4)

The primary funding decision at the headquarters level is how to divide the proposed total OST budget among Focus Areas, Crosscutting Programs, and other OST program units. Using budget guidance from Congress on a target figure for an overall OST budget, headquarters management proposes a budget for each program unit. The output of this proposal is an allocation that forms part of the executive Branch budget request to Congress. The OST budget is submitted by the executive branch and obtained through the congressional appropriations process, usually after some changes. Congress then approves the budget for each major DOE-EM program unit in the final energy and water appropriations bills for a federal fiscal year. Funds for RD&D activities within other EM offices (i.e., EM-30, 40, and 60) are part of the much larger budgets of these offices.

OST Headquarters Prioritization Process

Near the end of the committee's information-gathering phase, OST submitted a written description (DOE, 1997i) of and presented (Walker, 1997b; Barainca, 1998) a prioritization process that was in the final stages of development. This process would be used by OST to create an integrated priority list of work packages[4] at the headquarters level based on prioritized lists created and submitted by each program unit of OST (e.g., Focus Areas and Crosscutting Programs).

In the new prioritization process, OST personnel produce a numerical score for each work package. Weights are provided by "ranking factors," and scores are assessed using the scoring criteria within each ranking factor (see Table 4.2). These assessments use data submitted by the site with the work package and data from the database underlying paths to closure (DOE, 1998a). Each work package is evaluated against each of the scoring criterion within each ranking factor, and these numerical evaluations are combined with the weight of the ranking factors to determine a quantitative score. The new process is stated to be based on multi-objective decision analysis methodology. Beyond this generality, the processes used to determine the scoring criteria, ranking factors, and weighting factors were not specified.

OST used this process to establish the priority order of most work packages for the internal review budget (IRB) during the spring of 1998. In this system, the evaluations against the scoring criteria and the values of the weighting factors used are opportunities to exercise management judgment. Some projects (e.g., the EMSP) could not be scored under the new system, however, and were assigned priorities based solely on management judgment applied directly. For the work packages subjected to the process, OST representatives noted that the database underlying the prioritization system was not yet at a point where it was sufficiently accurate to provide the sole basis for prioritization, and management judgments were used

In this prioritization process, the highest-scoring work packages are selected for OST funding. Work packages with successively lower scores are then funded until the sum of the funds associated with work package projects exhausts the total OST budget. The budget of each OST program unit (e.g., each Focus Area or Crosscutting Program) can also be determined as the sum of the funded work packages under its management. In principle, this prioritization process can therefore be used to set budget targets for OST program units. In the past, these budget targets were set in a more subjective way by top-level OST management, taking into consideration EM strategic goals (such as risk or mortgage reduction; see Chapter 2).

TABLE 4.2 Five Criteria Used by OST to Rank Work Packages in FY 1998.

Ranking Factor	Scoring Criteria
Site needs	Need priority, which includes consideration of need timing
Project visibility	High visibility, which includes consideration of urgent or high risk and high cost
Future technology deployments	Planned for deployment, which includes consideration of deployment commitments
Closure technical Risks	Critical path to closure risk, which includes consideration of waste stream technical risk
Technology cost savings	Cost-saving potential

[4]A work package is a group of related projects, which in FY 1998, had annual costs ranging from less than a million dollars to tens of millions of dollars. In FY 1999, examples show the range to be from $0.2 million to $10 million, with an average of around $2 million. The total of these packages accounted for only about a third of the OST budget.

Allocations of Funds to Specific Technology Needs and Projects (Box 8)

The allocation of funds to projects within Focus Areas, Crosscutting Programs, and the Industry Program is driven by the program unit's prioritized lists of technology needs and projects. The highest-priority items within each program are funded, going down the list, until the budget is exhausted.

Solicitation of RD&D Proposals and Select Providers (Box 9)

The highest-ranking needs form the basis for soliciting proposals from industry, universities, and national labs, via appropriate solicitation announcements, and for continuation of worthwhile multiyear projects. Proposals received in response to these solicitations are evaluated against relevant technical and programmatic criteria, generally by technical review teams comprised of non-OST personnel, as well as OST program managers (see Appendixes C-E for further detail on the review procedures used). The highest-rated proposals, scored against both technical and programmatic criteria, are selected for funding.

Performance of Technology Development Projects (Box 10)

Projects selected for funding by OST are carried out by their proposers, who are PIs in national laboratories, universities, and private industry. Each funded project is tracked using performance milestones, technical status reports, and spending records. The goal of some Crosscutting Program projects is to transfer technologies under development to Focus Areas that can then support the remainder of the development work needed for eventual successful demonstration and deployment.

Project Monitoring and Evaluation (Box 11)

A major OST initiative to evaluate and monitor its projects was the adoption of the "stage-and-gate" model. Other efforts were to employ peer review (NRC, 1997b) and to adopt a suite of performance indicators to aid in the fiscal and technical management of technology development projects (DOE, 1997f).

In the stage-and-gate model (Paladino and Fox, undated; Paladino and Longworth, 1995) for mapping and tracking the maturity of a technology development effort, a project progresses from the basic research stage (stage 1), through bench-scale testing, applied engineering and development, and pilot- and full-scale demonstrations; to a final (seventh) stage, implementation by the end user (see Figure 4.2). Each step in the gate process indicates a level of maturation of the technology and an OST decision point to continue its funding support.

The stage-and-gate model has been refined (DOE, 1996o) with the development of gate criteria specific to OST and a formal OST gate review process. Once a project is funded, it is tracked in the stage-and-gate system, where gate criteria serve as decision points to promote the technology development project to a higher stage (with a correspondingly higher funding level and higher level of programmatic commitment). Gate and peer reviews on works in progress are used in principle for this purpose. The application of stage-and-gate and peer review enhancements to OST decision making is discussed in more detail in Chapter 5.

Gate 4 is an important gate, because it is the point of transition from a small-scale project to a significant, applied engineering development effort, requiring the commitment of significant resources to run larger-scale tests and demonstrations. OST recognizes that these expenditures should not be conducted without a review of the project's likelihood of success—meaning, in addition to technical success, the likelihood of its being used in DOE-EM and the likelihood of its being made into a commercially viable

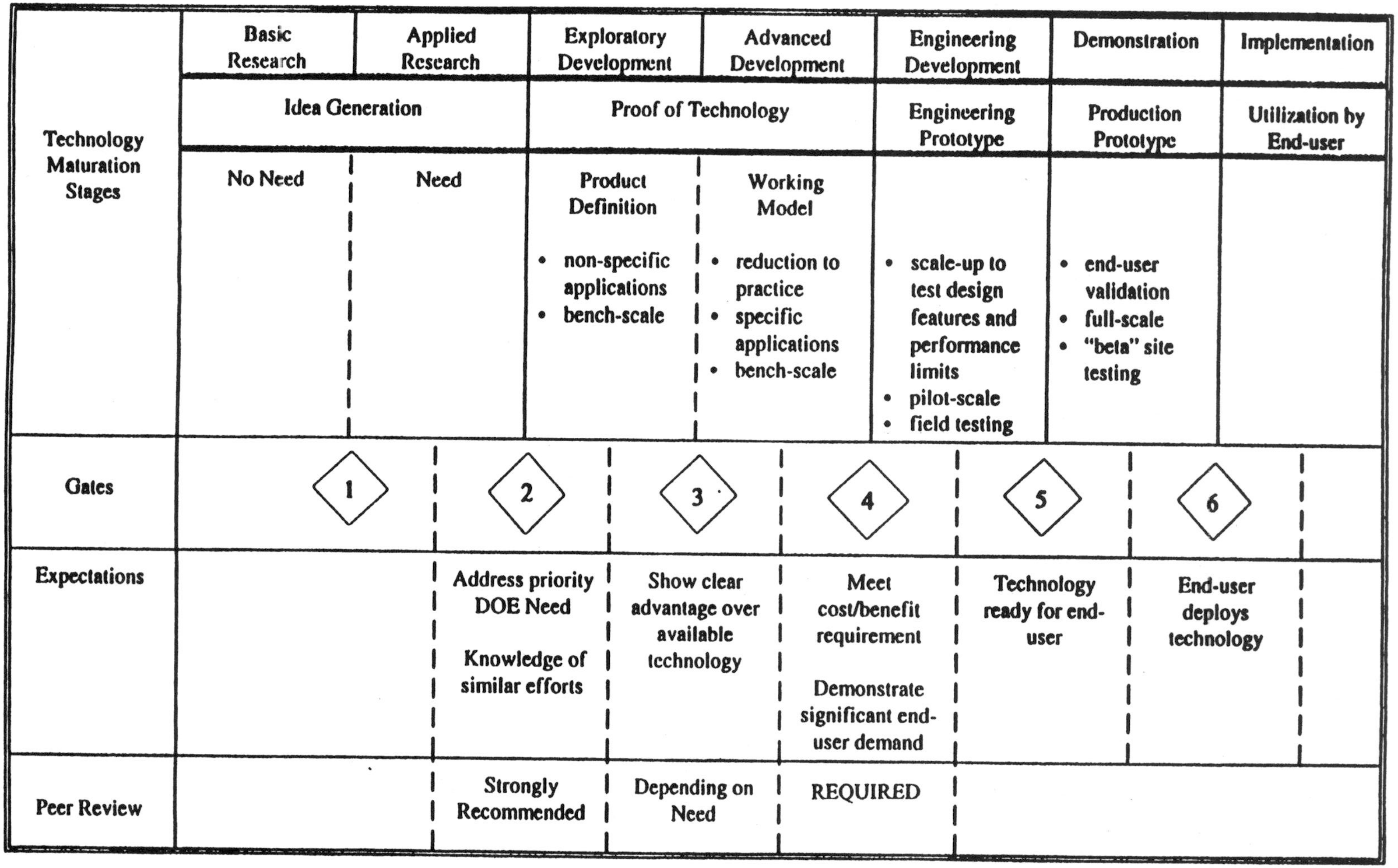

Technology Maturation Stages	Basic Research	Applied Research	Exploratory Development	Advanced Development	Engineering Development	Demonstration	Implementation
	Idea Generation		Proof of Technology		Engineering Prototype	Production Prototype	Utilization by End-user
	No Need	Need	Product Definition	Working Model			
			• non-specific applications • bench-scale	• reduction to practice • specific applications • bench-scale	• scale-up to test design features and performance limits • pilot-scale field testing	• end-user validation • full-scale • "beta" site testing	
Gates	◇ 1	◇ 2	◇ 3	◇ 4	◇ 5	◇ 6	
Expectations		Address priority DOE Need Knowledge of similar efforts	Show clear advantage over available technology	Meet cost/benefit requirement Demonstrate significant end-user demand	Technology ready for end-user	End-user deploys technology	
Peer Review		Strongly Recommended	Depending on Need	REQUIRED			

FIGURE 4.2. OST technology decision process. SOURCE: NRC, 1998b.

product. Such considerations are part of the gate 4 review criteria. For example, one criterion to pass through gate 4 is that the PI of a technology development project involve a private-sector business partner who is in a position to commercialize the technique. Another criterion is to achieve a DOE-EM field site "buy-in" to use the technology, if it meets its advertised specifications.

Technology Deployment (Box 12)

During 1997, the historically slow pace of technology deployment caused OST to propose and Congress to fund a program initially called the Technology Deployment Initiative and later renamed the Accelerated Site Technology Deployment Program (described in Chapter 2 and Appendix E). The purpose of this program was to facilitate initial deployment of technologies developed by OST or the private sector by providing funding specifically for this purpose. A number of conditions were attached to this funding, including the need for problem owner endorsement and cofunding, and the identification of likely follow-on deployments. The initial set of ASTD projects was funded in FY 1998.

CHAPTER SUMMARY: MAJOR DECISION PROCESS STEPS AND HOW THEY WERE DONE WITHIN OST DURING FY97-98

An important context for OST decision making is provided by the following features:

- the annual federal budget system and other oversight requirements, which drive each EM office to plan for several fiscal years into the future during any one fiscal year;
- the nature of funding for the OST program, with the budget not directly connected to immediate DOE-EM activities undertaken in support of compliance and cleanup agreements; and
- the fact that other EM offices (shown as user programs in Figure 4.1) are the ultimate customers of technology at EM sites and control both the baseline remediation planning (Box 5) and the deployment of technology (Box 12).

The decisions at these levels impact OST's program, yet they are largely outside the direct control of technology developers.

Given this context and a functional description of the decisions that must be made, a general decision-making process can be formulated (Figure 4.1) to describe the current system. The process steps of Figure 4.1 are carried out with input from various elements of the federal government, including Congress, OMB, the office of the Secretary of Energy, DOE-EM problem owners, DOE-EM technology developers, and site-specific stakeholders such as states, regional EPAs, Native American tribes, and the interested public.

Figure 4.1 depicts the external constraints on OST (in the top and left-hand sections) and the internal process that OST is attempting to implement (in the right-hand section). This figure can be used in several ways. One way, pursued in this chapter, is to offer a description of how OST accomplishes the functional steps represented by individual boxes. Another way, pursued in Chapter 5, is to offer evaluative comments on how well these OST methods perform the requisite functions. Still a third use of this figure is to raise issues such as whether the diagrammed process is viable or in need of additional functional steps, issues also considered in Chapters 5 and 6. The figure raises yet another question of whether radically different processes might be constructed to be effective. No finding or recommendation is offered on this last issue, insofar as the committee's focus was on assessing and improving the needs-based approach that OST is pursuing.

The history of OST presented in Chapter 2 and the context, function, and process steps discussed in this chapter lead to findings and recommendations in Chapter 6. In developing these, Chapter 2 has looked at the decision-making practices of private industry to gather additional information and insight. The results of these assessments are presented in Chapter 5.

5

Application of Good Practices in Prioritization and Decision-Making Processes Relevant to OST

The first part of this chapter discusses the general themes introduced in Chapter 3 in the context of the decision-making environment described in Chapter 4. This treatment presents ways in which these themes can find application in OST, organized by headings that highlight some of the relevant best practices.[1] Following these comments of a general nature, more specific evaluations are provided in the second part of this chapter on OST prioritization and decision process steps. These are evaluative comments on the ways in which the functional process steps of Figure 4.1 are conducted within the current OST program. Findings and recommendations arising from each of these parts are contained in Chapter 6.

The committee has reviewed the prioritization and decision-making practices of OST (Chapter 4) in the context of the best practices introduced in Chapter 3. The results of this comparison are presented in this chapter and form the basis for many of the findings and recommendations in Chapter 6. Discussion of OST's decision-making process is organized by the sections that follow, which highlight some of the relevant best practices. Following these comments of a general nature, more specific evaluations are provided of OST process steps using the model of Figure 4.1.

AGREE ON CLEAR AND MEASURABLE GOALS

A most important requirement for good decision making is the establishment by the top levels of management of an organization's strategic goals (see Simon, 1997). The goals must be broad and clear enough to provide guidance to all parts of the enterprise, yet specific enough to provide a basis for effective decision making at all levels of the organization. This requires definition of a hierarchy of goals and responsibilities for each organizational component and program. In addition, it should be possible to measure how well these goals are met. Although former Assistant Secretary of EM Thomas Grumbly established consideration of "risk" as a major strategic goal for EM, and former Assistant Secretary Alvin Alm established "reducing the mortgage" as his major strategic goal, neither of these goals was specific enough to allow OST (or other EM decision makers) to differentiate effectively between short-term and longer-term priorities. As a result, very important, but long lead-time problems, could be relegated to priorities so low that they were severely cut in funding, even to the point of not appearing above the

---

[1]The only one of the 10 practices highlighted in Chapter 3 that is not discussed here is "hire the best people possible and maintain expertise," because OST is constrained by DOE and federal government hiring practices that are largely outside OST's direct control. Some program managers interviewed for this study (see Appendixes C and D) were no longer in the OST program by the time this report was published. Committee thoughts on how to apply this idea to OST are mentioned in Chapter 3.

budget-constrained cut-off line for funding. Publication of DOE-EM's strategic plan (DOE, 1997j) and *Accelerating Cleanup: Paths to Closure* (DOE, 1998a; 1998b) helps to address this prior lack of specificity.

It does not appear to the committee that OST has established a uniform set of quantifiable, measurable goals and criteria needed for high-quality decision making. These top-down goals for OST can be derived from user input such as "strategic goals" (DOE, 1997j), which target the sum of the work to be achieved at all sites in a bottom-up approach. These strategic goals are obtained through a user-driven and, therefore, bottom-up decision-making process of other EM offices in which most remediation decisions are made at individual sites. Top-level goals derived from these and other user inputs (e.g., DOE, 1998a; 1998b) could be used in a centralized, coordinated, top-down process within OST to drive the technology development program. Of course, any top-level strategic goals developed by OST must be consistent with the EM mission and derived in coordination with user plans and needs.

UNDERSTAND AND FOCUS ON CUSTOMER NEEDS AND REQUIREMENTS

As OST learned by hard experience, the technologies it has developed are of value in cleanup only if the intended technology users accept and deploy them. The fact that a technology may be used by industry for tasks other than site cleanup, although important to the developer, is not an essential criterion of the user of a technology in DOE-EM. As the level of acceptance of OST technologies by the sites came under increasing criticism, OST employed several different approaches in its attempts to get new technologies deployed at the sites. The most recent attempts are embodied in the ASTD approach, which provides cofunding to the sites as an incentive to deploy technologies, and in the TAC and other headquarters integration activities intended to facilitate deployment using upper DOE management interactions. Whether or not these approaches will be successful has yet to be determined. In any event, OST has attempted to overcome the technology deployment obstacles it faces.

COMMUNICATE ACROSS ORGANIZATIONAL BOUNDARIES

Interaction, with frequent feedback, between OST managers at the sites (i.e., managers of Focus Areas and Crosscutting Programs) and site contractors and their subcontractors is highly desirable. By maintaining a continuing awareness of the sites' evolving technology needs and an understanding of the many constraints and pressures that site operators face, OST may increase its effectiveness in substituting cheaper and/or more effective technologies in the baseline functional flowsheets.

THINK STRATEGICALLY:
HEDGE AGAINST TECHNICAL UNCERTAINTY AND INSIST ON ALTERNATIVES

A much more difficult problem than technology deployment is obtaining broad recognition that alternatives to the baseline functional flowsheet are an essential part of sound systems engineering and that good decision-making practices dictate that all well-planned and executed projects allow for the possibility of failure and provide for contingencies in this event. The constrained budgets for cleanup, combined with pressures to move more quickly in cleaning up sites, have forced decision makers to eliminate any significant consideration of alternatives. Yet experience shows repeatedly that failure in RD&D efforts does occur. The good decision-making practices discussed in Chapter 3 also highlight the need to consider alternatives (see also Simon, 1997), which is the responsibility of the community of users of technology at DOE-EM sites.

Recognition of the need for alternative solutions to cleanup problems in the event of technical failure or funding shortfall is essential. Alternative functional flowsheets or overall problem solutions often require that alternative technologies be developed.

In a similar vein, it is appropriate for OST to have input in the initial establishment of baseline functional flowsheets, even though the problem owner at the site bears the responsibility for cleanup. Because some of the most intractable of DOE's cleanup problems are the first of a kind in many respects, it is unlikely that any site contractor will be totally prepared for all of the problems to be faced.

OST and its established contracted technology experts are major potential resources of technical expertise for baseline functional flowsheet discussion and evaluation to the winning bidders of any privatization contract. Through the use of its available expertise, OST could provide needed assurance to DOE-EM that the contracted cleanup has a better likelihood of success. It may not be sufficient for DOE to take the position that the entire cleanup responsibility lies with the site contractor, even though this may be what the contract calls for. Such a course puts the taxpayer at risk; the taxpayer must pay for failure in terms of lost time, continued risk, and additional cost.

Application: Balance Innovation and Incremental RD&D

The process of structured, thorough screening and decision "gates" is appropriate for most development work on behalf of DOE-EM (this applies to work conducted by OST and other EM offices). Similar structured processes are generally used by industry counterparts, with emphasis on cost-benefit evaluation of the expected results of development work.

Although this report strongly endorses more structured evaluations, too rigid an application of screening criteria—at least in the early stages of some types of RD&D—can lead to a portfolio of RD&D efforts that becomes unbalanced by an overemphasis on projects and tasks that fit into short time horizons (i.e., with measurable payoffs in one to three years). This practice, particularly of funding only short-term, late-stage projects, would tend to exclude any project involving substantial innovation, that is, any project with technical risks, particularly if the proposed project is not fully understood or based on well-known science and technology. Innovation may be seriously discouraged if there is not some policy of acceptance of reasonable technical risk. Success rates for industrial RD&D depend on the mix of more risky RD&D projects with less risky (but often necessary) design and development projects.

The committee believes that an essential need for the DOE-EM RD&D function is to maintain a *portion* of its development work in research projects that are not primarily refinements of existing technology. This means exploring significantly different concepts. The intent should be to maintain a hedge against the possibility that important parts of the baseline technologies may not work well enough or may prove to be much more costly than originally expected. This also recognizes that most truly innovative concepts do *not* pay off, yet many must be tried in order to find the occasional big winner. Therefore, decisions on what technology to fund should be made in the context of exploring new concepts *having direct relevance* to cleanup, however difficult such judgments may be. In the United States, researchers in university or other independent settings are considered the traditional sources of innovative ideas. However, on-site technical people are most familiar with process limitations, and they should also be canvassed regularly for ideas.

The industrial RD&D programs, as well as those of nonprofit organizations such as EPRI and GRI, generally recognize the need to maintain at least a small "exploratory" or "strategic" RD&D category. This is commonly budgeted in the range of 10 to 12 percent of overall RD&D expenditures. This "hedging by exploring innovative ideas" is widely recognized as a good practice for the prudent management of RD&D portfolios.

CONTINUALLY IMPROVE THE RD&D MANAGEMENT PROCESS

In cases where the need is great and the outcome of a technology development project to meet the need is uncertain, it is justifiable to fund some overlapping or duplicative technology development projects, especially at the early, less expensive stages of development. However, in the past, OST has supported the demonstration of already proven and very similar technologies (GAO, 1996; see also Appendix C, Boxes C.1 and C.3). Partly as a result of budget cuts, partly as a result of increased sophistication on the part of OST managers, and partly as a result of the maturation of cleanup activities and concomitant focusing on the most important needs, this duplication of technology development activities by OST has likely been reduced.

Another area of duplication of effort by OST is in the acquisition of commercially mature technologies from industry. This activity appears to be carried out by the Industry Program at the Federal Energy Technology Center (FETC), by the DDFA (via the LSDPs; see Appendix C), and by site operators. (Focus Areas often use their own budgets for internal-to-DOE-system procurements, for example, to national labs.) Although there is some coordination of procurement based on needs formulated by Focus Areas, the practice of independent funding for procurement tends to undercut effective coordination. The most common OST practice is to put a statement of technology need in the *Commerce Business Daily* and wait for responses from industry. A preferred approach would not only consider the availability of technologies in industry, but also provide an assessment of the strengths and shortcomings of these technologies and their effective integration into site functional flowsheets before any technology development solicitation is made.

OST has recently begun to use the U.S. Army Corps of Engineers to carry out cost studies of technologies developed, or proposed for development, by OST and to compare these costs with those of the baseline functional flowsheet technologies they are intended to replace. This is another attempt by OST to show top-level managers of DOE-EM, members of Congress, and "watchdog" groups that OST is performing a needed and responsible job. The quantification of problems by this approach is a way to generate informational inputs to a decision process. However, there is the ever-present problem that the cost of the baseline flowsheet is often not known to an acceptable degree of certainty. For credible cost comparisons, calculations of baseline and alternative technologies should be done in a consistent way (NRC, 1998a).

A further addition to OST's attempts to be responsive to criticisms of its productivity, as measured by deployment of the results of its efforts, is the introduction of a comprehensive database that includes technologies it has developed, their costs, and whether or not they have been deployed. This database allows OST to be responsive to various watchdog groups in a timely way and provides a measure of OST's deployment successes. It also provides an effective tool for use by OST in cost-benefit determinations, which provides valuable decision-making information.

An inherent weakness in any attempt to assess the value of RD&D is the fact that it takes time for technologies to be demonstrated and deployed. Consequently, many of OST's technologies not yet credited with being "successes" may be successes in the future.

Application: Document Bases for Retrospective Analysis, Especially in Headquarters' Discretionary Allocations

The OST program management at headquarters in the past has funded some projects directly, without using established review procedures as part of its decision making. Flexibility to act in this way is appropriate because it provides a way to respond quickly to a late-breaking development or opportunity. An example of this method of decision making was the OST headquarters decision to fund the Hanford Tank Initiative (HTI), which the Hanford STCG had endorsed.

As pointed out in Chapter 3, it is good practice to document the basis for making decisions, especially those that are made outside the established process. If a structured decision-making process is not followed, documentation to allow retrospective examination of discretionary decisions seems prudent so that lessons can be learned from past decisions and adverse reactions can be avoided.

MEASURE AND EVALUATE TO GUIDE RESOURCE ALLOCATION

The STCGs and Focus Areas use criteria for prioritization of needs and selection of projects. These criteria were formulated in an attempt to provide a rational basis on which to select and prioritize the technology needs forwarded to OST by problem owners at the sites. However, there has not been a concerted effort to make these criteria uniform (to the extent practicable) from one STCG or Focus Area to another (see Appendixes B and C). Furthermore, many of the criteria that do exist are not well formulated. They contain both obvious and underlying redundancies and in some cases are cast in such a way as to permit important, and even vital, needs to be downgraded so far that they are dropped from consideration. Guidance is needed from upper levels of DOE management for that development of formal, consistent, and rational criteria across all of OST and for "threshold" levels for each criterion to ensure that vital needs are met.

Application: Return-on-Investment (ROI) Evaluations for OST

ROI calculations were done in the 1997 ASTD program of OST. The use of ROI calculations to evaluate ASTD proposals is a practice that could be extended by applying it to all technology development project proposals prior to their final selection for funding, both within each Focus Area (and Crosscutting Program) and across these program units, to identify the areas of greatest return on technology development investments. Returns could include benefits in many areas, as measured by estimated improvements in cost, schedule, safety, risk, and/or cleanup levels. ROI-type evaluations would also provide information by which to compare projects of separate program units (i.e., the relative worthiness of a Tanks Focus Area project versus one from the DDFA, or from the Robotics Crosscutting Program), thereby helping to inform decisions of how much money should be allocated to each program unit.

In 1997, OST used the Army Corps of Engineers to provide an ROI calculation for the entire OST program (DOE, 1997h; U.S. Army Corps of Engineers, 1997). The calculation estimated financial savings in the form of a "cost avoidance" to the federal EM budget. The Corps used a subset of OST technology development projects and extrapolated the deployments that could be achieved from these projects to derive an overall ROI for the $2.6 billion spent by OST since 1989, if the OST-developed technologies were deployed to replace some of the currently conceived baseline remediation technologies. Deployment assumptions introduced uncertainty into the calculations that was treated by calculating a range of financial returns, based on a range of assumptions about how aggressively these technologies would be deployed at sites.

Unfortunately, many of these calculations suffer from the limited credibility of the baseline cost estimates.[2] These cost comparisons cannot be done at all if a baseline remediation plan is lacking (because there is no baseline cost and technology to compare against). Another problem of the credibility of the estimated baseline cost savings when a baseline remediation technology is excessively costly; if the baseline

---

[2]More discussion of baseline cost methodology is found in NRC (1998a).

remediation approach costs too much to be a realistic approach, then savings relative to baseline cost figure are not true, credible savings. However, despite these problems, such estimations are worth considering because a baseline plan must be developed and credible costs must be generated or else the cleanup may never move forward.

Application: Measures to Capture Technical, "Market", and Cost Information

Evaluation of an RD&D project can be done against many measures, such as the projected (1) benefits of using the technology for a particular application; (2) opportunities (i.e., other potential applications) that the technology offers; (3) costs incurred in its adaptation and deployment; and (4) risks of this deployment. A decision to deploy a newly developed technology could be made based on positive projections in any one (or more) of these four areas. Hence, one possible composite measure might be the sum of the four, as measured in a common unit (e.g., an avoided cost).

Alternatively, another measure can be constructed based on the concept that success at technology development requires success at both technical development and deployment. Thus, evaluation of a technology development proposal should include the combined probabilities of technical and deployment success. The potential benefit of a technology development initiative could be expressed as the product of these probabilities with the cost savings of performing a job using a new (versus an already established) technology. This benefit should be weighed against the development costs. This concept can be expressed mathematically by the following equation:

$$\text{Figure of Merit} = (P_T \bullet P_D \bullet \Delta_{cost})/(C_D),$$

Where
P_T = probability of technical success (e.g., GRI makes this proportional to the gate number in the stage-and-gate model);
P_D = probability of market penetration (i.e., potential for deployment on site applications);
Δ_{cost} = difference in cost of doing a job in an old versus a new way (i.e., the old cost minus the new cost); and
C_D = cost of developing the technology (i.e., performing the RD&D).

This formula expresses a benefit-cost ratio that could be a useful measure for evaluating the worthiness of technology development proposals and ranking them. This formula illustrates the basic factors in a simplified way.

In the prioritization and selection of technologies for development, other factors may also be treated probabilistically. These include the probability of solving significant DOE-EM problems (which may be, but are not necessarily, proportional to the number of sites having problems addressed by the technology under consideration) and the probability of funding (which may be related to the benefits as measured against risk, cost, and mortgage reduction). Another probabilistic factor is timeliness, insofar as the probability of sustaining adequate funding for a technology diminishes as the need to solve the problem addressed by the technology becomes more distant in time. These probabilities might be obtained initially from "back-of-the-envelope"-type estimations or judgments of knowledgeable people, to be supplanted over time by more refined figures as warranted, as data become available, and as the maturity of a technology development project increases. Despite all of the problems associated with these quantitative estimations, they are recommended to provide guidance because the alternative (of not doing them) provides less guidance to decision making.

The depth of any benefit-cost evaluation should be commensurate with both the proposed technology's development cost and the available knowledge. Although a simple analysis may suffice initially, a more refined, sophisticated analysis may be needed later on. In rigorous application, it is critical to determine the present worth of future savings using a well reasoned and documented value for the discount factor or factors used. For such calculations, credible input data are needed for reliable estimates of cost, risk, and schedule. If credible data are not available, then it is not feasible to gauge the probability of project success with this approach.

Application: Review Resource Allocations and Inputs

In the present situation of technology and program reviews, a good case can be made that OST (and, indeed, DOE-EM) is subject to burdensome reviews. This does not mean that there should not be reviews. However, an excessive number of reviews consumes the time and energy of OST staff; the number should be reduced. This could be done by appropriately combining some reviews and deleting others, based on an analysis of the objectives to be served by these reviews. In addition, there does not appear to be a need for a large attendance at the reviews. Careful selection of key reviewers, both OST staff and outside consultants, could substantially reduce the cost in time and money spent on reviews.

Although the reviews depicted in Figure 5.1 serve a variety of purposes for OST, most of them are not independent, expert, external technical "peer reviews" as defined narrowly in NRC, 1997b. For this reason, even though the committee has said that there is an excess of reviews of OST, the committee suggests that, in two very important places, additional external, independent reviews can benefit the OST decision process depicted in Figure 4.1. These reviews, discussed below, are to validate the following two inputs to the decisions made at Box 8 of Figure 4.1:

1. allocation of funds among problem areas at the policy level, made by headquarters management, and
2. process of identifying baseline processes, technology contained in the baseline functional flowsheet, and the derived technology needs, developed by the sites.

These reviews are considered below in more detail. The second can be conducted in a way to satisfy the requirements of a true peer review (NRC, 1997b), and presents an opportunity for knowledgeable OST contractors to access important information if the reviews are structured to include their participation.

External, Independent Review at the Policy Level of the Allocation of Funds

The decision about how to allocate OST funds among broad "problem areas" (e.g., Focus Areas) is a major step in the strategy to commit funds to the greatest areas of technology need for DOE-EM site cleanups. A review of the extent to which this allocation is consistent with DOE-EM strategic direction, priorities, and program goals by an independent, expert group having no stake in the allocation would be of value to top program managers. A logical body to perform this review might be a group outside OST, but familiar with the program, such as the Environmental Management Advisory Board. This board includes representatives of the public, private, and regulatory sectors and could therefore provide a particularly valuable, broad perspective on OST's program.

Peer Review of Baseline Processes and Technology Needs Identified by the Sites

The technology needs for a DOE-EM site are derived from the baseline functional flowsheets and alternatives, as recommended herein. Thus, these flowsheets are of fundamental importance to cleanup success and to decisions OST makes on technology RD&D funding. A peer review of the baseline functional flowsheets could serve three very useful purposes:

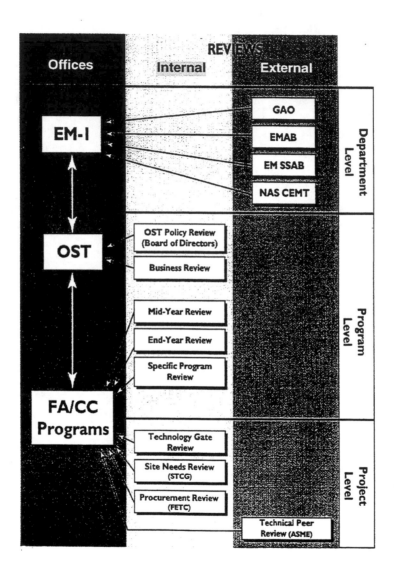

FIGURE 5.1   Diagram of OST's review program showing different types of reviews, offices to which reviews are submitted, and the level of the organization of which reviews occur.
NOTE: EM-1 = Office of Assistant Secretary for Environmental Management; ER = Office of Energy Research.
SOURCE: NRC, 1998b.

1. It could help identify those needs that could be met by technologies that are already commercially available in the private sector, in the U.S. or abroad.

2. It could identify opportunities to develop technologies that could significantly improve on the baseline approaches or could provide a fallback option in case there is some risk that the baseline process will not work as expected.[3] Availability of technical options that can provide adequate performance at substantially reduced cost could be particularly important if constrained funding threatens the ability to proceed with timely cleanup using the baseline technology (NRC, 1996b).

3. It could bring a broad range of technical expertise and experience to bear on difficult or presumed intractable technical problems.

OST's access to the results of such peer reviews could enhance its interaction with its customers. Indeed, OST's technical contractors at other sites might be considered as peer reviewers of a site baseline functional flowsheet, if their technical expertise, external viewpoint, independence from the work being reviewed, and objectivity can all be established.

SUMMARY AND CONCLUSIONS—GENERAL ISSUES

The OST program structure of FY 1997-1998 in principle, can perform all essential functions of RD&D decision making. However, improvements are needed to establish a formal, documented, and more quantifiable decision-making process. Improvements to the current program organization can come by instituting

*   structured and effective information flow between OST program units;
*   documentation of the bases for discretionary allocations (e.g., funding decisions from headquarters that are outside the decision-making processes of the Focus Areas, Crosscutting Programs, and other OST program units described in Appendix E);
*   appropriately targeted independent, external reviews (or "technical audits"), including a review of site baseline functional flowsheets generated by site problem owners and from which technology needs statements are derived; and
*   support of significantly different concepts *directly relevant to cleanup* that may come from sources other than user-generated statements of technology needs. One common way to do this (as at EPRI and GRI) is by supporting exploratory research using a fixed percentage of the program budget.

---

[3]The EMAB Technology Development and Transfer Committee identified the need for improvement and expansion of the technology development process to include technologies that could improve on the baseline (Berkey, 1997). Along the same lines, the NRC Committee on Buried and Tank Waste, in its review of the Environmental Impact Statement for the Hanford Tank Waste Remediation System, concluded: "Backup approaches [to the baseline] are needed because the technologies projected to meet current requirements might not work or might cost far more than anticipated . . . DOE should develop fallback options and promising alternatives that might achieve most of the projected benefits of current options at a substantially reduced cost (NRC, 1996c, p. 48)."

USE OF A FORMAL DECISION-MAKING PROCESS: EVALUATIONS OF DOE-EM'S IMPLEMENTATION OF THE PROCESS STEPS OF FIGURE 4.1

This section provides a step-by-step evaluation of the decision processes presented to the committee during its information-gathering meetings, following the box-by-box diagram of Figure 4.1. A summary of the information presented during these meetings is included in Appendixes B-E.

User Program Funding (Boxes 1 and 2)

Stated briefly and somewhat oversimplified, budget requests from each user program (e.g., programs at various DOE sites) are compiled based on general guidance from the upper levels of DOE management. The budget requests are then aggregated at DOE headquarters. Some further adjustments among the major components of DOE typically occur at this stage, based on judgments from the Office of the Secretary. Such adjustments are frequently required to meet overall budget goals. This compiled request goes to the OMB. The OMB may make or recommend further adjustments in respect to both the overall DOE budget request and the balance among major DOE organizations. The OMB does not usually recommend adjustment of the components of user programs, although new initiatives are an exception to this. The administration then submits the budget request for consideration by Congress via the applicable congressional appropriations committee.

Congress may approve the budget as requested or, more commonly, make further changes based on the input of the appropriations committee. These adjustments are subject to some dialogue at each stage between the adjusting body and the relevant management levels in DOE. The budget eventually passed by Congress is then made available to DOE, which can issue authorizations for expenditures and make changes. Sometimes, portions are withheld at the discretion of the administration, subject to allocation later in the fiscal year.

Most of the budget request and approval process outlined above occurs two fiscal years before expenditures are authorized and implemented. If budgets are reasonably stable (i.e., predictable within an uncertainty of approximately 5 to 10 percent), the two-year advance pattern has the advantage that orderly planning and execution of programs is possible on a time scale of two or more years. However, this is often not the case. In particular, there are instances in which the budget for the next fiscal year is substantially different than the budget for the fiscal year after that. In addition, events occurring during a year can cause out-of-cycle budget changes (usually decreases). One example is the extensive flooding in the upper Midwest in the winter-spring of 1996-1997, which resulted in massive federal financial aid that, in turn, manifested itself as funding reductions (called rescissions) during the middle of the year. Although many natural events cannot be anticipated, budget changes shortly before or during the year of expenditure can be particularly destructive because they may redirect or terminate user projects that form the basis for the user program's technology needs. This leaves decision makers in the technology provider program with the difficult choice of either terminating a technology development project, with zero return on the investment to date, or continuing the project, with the expectation that it may be needed later. In addition, this reopens some of the decisions on the relative emphasis on different goals and measures of success of the agency, of EM, and of OST and therefore on the programs mounted to accomplish these goals.

This budget process has the following important implications for the decision-making process:

- The OST budget must fit within the boundary conditions of higher-level budgets. In the present climate, this usually restricts the allowable budget to much less than what would be required to fund all of DOE-EM's science and technology needs.

- The DOE is left with a choice between two less-than-optimal alternatives:

    1. plan a bottom-up budget based on focus area input two years before the program execution year in an environment that changes much more frequently, or
    2. submit a generic budget and develop the detailed work plan in parallel with final approval of the budget.

DOE has chosen the latter alternative. Given the boundary conditions and the importance of being relevant to user needs, this choice is reasonable. However, it is important to note that this means the OST budget negotiated with OMB and the U.S. Congress is independent of the detailed program formulation by the Focus Areas, which must necessarily proceed in parallel and be adjusted as the negotiations change their allocation.

- The decision-making process must allow for ongoing adjustments during program execution to accommodate both technological and budget changes.

Funding for the Technology Provider Program (Box 3)

The process for allocating funding to the technology provider programs (i.e., the program units of OST) is essentially the same as that described above for the user programs, primarily because both are part of the same DOE-EM budgetary process. Decisions on top-level OST budget requests and adjustments are apparently made by upper-level OST managers.

The discretionary nature of OST program funding makes it particularly vulnerable to "directed funding" (also known as "earmarks"). These are directives communicated via congressional budget legislation and other less formal mechanisms to fund specific projects, often at specific organizations. These effectively bypass the decision-making process shown in Figure 4.1 by requiring that certain activities be funded irrespective of their merit. Although Congress has the right and obligation to establish, terminate, or alter programs, the committee believes that its intervention in what are essentially decisions at the project level constitutes micromanagement that compromises the integrity of the decision-making process. Other high-level review committees (PCAST, 1997) have recently reached similar conclusions. The committee urges that the OST program be allowed to establish and execute a transparent, high-integrity, decision-making process that includes all of its activities at the project level, and that upper-level management and budget authorities exercise their responsibilities at the higher program and policy levels.

Funding Allocations to OST Program Units (Box 4)

In the past, the Board of Directors (BOD) considered information from the Focus Areas to determine their budgets. The committee's only information regarding BOD methodology came from discussion during a March 1997 committee meeting with Dr. Clyde Frank, then Deputy Assistant Secretary for Science and Technology. In principle, the information about alternative technology development projects to fund, and prioritizations within each Focus Area, should be available for determining each Focus Area's budget. With this information in hand, the BOD could gauge the trade-offs that have to be made when cutting or increasing funding levels in a given Focus Area. During the BOD allocation process, this information appears to have been either unavailable or unused at the level of detail necessary to make sufficiently informed choices. As a consequence, the BOD appears to have made allocation decisions largely based on very general considerations and trade-offs, and not on specific priorities of projects or technology needs.

Adjustments to these funding levels are made in the Program Execution Guidance (PEG) annual review meeting held at the end of each fiscal year. This is an internal review by OST headquarters management of all major OST program units (e.g., Focus Areas and Crosscutting Programs), based on presentations by the units' senior program managers. The value of the total OST budget for the next fiscal year that is received from the congressional appropriations process (Box 2) is considered against each program unit's funding request, to adjust each program unit's budget target for the upcoming fiscal year. The scopes of work are reviewed to explore cost-cutting opportunities and the relative worth of various projects. In FY 1998, this budget allocation step was treated to an analysis provided by a decision support tool, the "new prioritization process" described in Chapter 4 and discussed next.

Evaluation of the OST Headquarters Prioritization Process

It was not possible for the committee to fully evaluate OST's 1998 headquarters process (Barainca, 1998; DOE, 1997i; Walker, 1997) for prioritizing work packages[4] at the headquarters level because this process was not yet complete. However, based on the information presented in Chapter 4, the committee has a number of concerns about the proposed process that are offered below to provide guidance for the final stages of its development.

There are inherent problems with prioritizing work packages composed of multiple needs or projects. In principle, the projects in a work package have differing cost, risk, and benefit attributes related to enabling or improving resolution of different DOE-EM problems. As a consequence, low-priority projects could be carried along with others having a high priority, or high-priority projects might not be funded because of a preponderance of low-priority projects in a work package.

The purpose of the process of prioritizing work packages was not stated and was not evident. The only potential purpose the committee could identify is to provide a basis for deciding which work should be terminated if the budget is cut. However, as noted immediately above, termination of a work package eliminates multiple projects, each of which has various priorities and levels of merit. For example, one work package involves developing waste retrieval technology for three sites, where the risks and priorities are likely to be different. Further, elimination of a work package could render an entire process infeasible. For example, if a retrieval technology is unavailable, this may preclude subsequent operations such as waste processing, immobilization, and disposal. Strict application of the prioritized list is likely to have unintended dysfunctional results that can be avoided only by discussing the priorities on a project-by-project basis with organizations at the sites (e.g., Focus Areas and problem owners). This leads the committee to question the value of the aggregated prioritization.

The database required to support a valid prioritization is not yet in place, a fact that was clearly stated by DOE. The committee believes it likely that the uncertainties in risk and cost data, when coupled with instability of the site-driven needs, will result in the uncertainties in the database being so great that the utility of the results is highly questionable. These uncertainties result from both inaccurate information (e.g., inability to estimate cost savings in the distant future) and differing judgments on qualitative criteria such as "high visibility."

---

[4]Work packages are aggregates of one or more technology needs. Insofar as ongoing projects for already-solicited work are in progress in any given year, needs identified in a previous budget year are represented by technology development projects in progress. Any work package can therefore have a combination of new starts (representing new needs) and work in progress (relevant to previously identified needs).

The committee was unable to determine how priorities imposed by OST would be reconciled with user-driven priorities established by site organizations such as STCGs and Focus Areas in the event that there are differences.

The committee did not have enough information on this evolving prioritization process to reach firm conclusions about it. However, the committee believes that OST should evaluate its current approach in the context of the above concerns. A peer review of the proposed prioritization approach, once it is fully formulated, could be valuable.

User or Site Program Planning (Box 5)

As discussed earlier, the technology user programs develop baseline functional flowsheets (and, hopefully, alternatives), from which technology needs are derived. The lack of formal, substantive OST involvement in this planning activity implies that OST cannot now influence the technical decisions from which are derived the technology development needs that OST gathers at each major DOE-EM site. This is a substantial deficiency in overall program efficiency.

Identification and Prioritization of Technology Needs (Box 6)

OST accomplished the steps in Box 6 using Site Technology Coordination Groups, which are discussed in Chapter 4 and Appendix B. In the 1994 EM-50 program reorganization, the STCGs were formed to strengthen the tie to the customer. The STCG's mission was to document and prioritize technology development needs at a DOE-EM site by using input from site managers and operating contractors for EM-30, 40, and 60. Each STCG forwarded a prioritized list of site needs annually to the Focus Areas, which gathered multiple site needs, established a national prioritization, and requested funding for work relevant to the highest-priority needs. In principle, the Focus Areas owe the STCGs a response for how each need is addressed in the suite of technology development projects selected for funding, but the committee was told informally that this feedback was often missing.

Although the STCG processes all differ, they have similar strengths and weaknesses. One strength is that the resulting lists of priority needs reflect a site's priorities both within a particular topical area (e.g., mixed waste, tank waste, subsurface contamination, or decontamination and decommissioning) and across topical areas (e.g., all site needs in EM technology development can be rank-ordered based on the STCG process). A possible weakness is that the technical needs of the site problem owners may not be accurately communicated, since the needs statements are often by design recast to be more general descriptors than those that are received from site contractors who own the problems (the user community).

All STCGs use a structured and documented process to identify needs. Typically, this process involves a series of meetings that generate annotated lists of technology needs and involve discussions and prioritization of these needs. A clear strength of a structured and documented process is that it provides an accounting trail for subsequent decision making and external review. It also allows for both technical input (at the subgroup level) and stakeholder input (at the higher STCG management level).

All STCGs attempt to use a formal prioritization methodology. Typically, these formal methodologies consist of defining sets of prioritization criteria, scoring technology needs on these criteria, weighting the criteria, and calculating some weighted average to indicate the overall priority of the technology needs. The strengths of these formal methodologies are that they provide a consistent framework for prioritization and provide an accounting trail and defense for prioritized lists of needs to be reviewed in higher-level decisions. However, the committee observed several weaknesses in the use of these formal methodologies. Many STCG criteria were poorly defined (with short phrases that were subject to different interpretations in the absence of more expanded descriptors) or were redundant (e.g., criteria of risk reduction and of safety benefit are somewhat redundant). Some methodologies used weights that could exclude essential needs. It is possible, for example, to exclude an essential need such as safety by giving it so little weight it does not affect the selection of a technology. As a result, the committee concludes that many of the prioritization criteria fall short of being truly effective and rigorous. Further, they differ from site to site, making it difficult to ensure consistency of results across sites.

In addition to being a resource to gather site needs, STCGs may suggest a demonstration site for a developed technology and provide nonbinding endorsements of technologies or proposals (e.g., the Hanford Tank Initiative was endorsed by the Hanford STCG) to OST management. In these efforts, the STCG facilitates the demonstration and implementation of technologies at its site, thereby serving as an advocate for that site in promoting constructive technology development activities, with both OST and the site perceived as winners credited with any successful venture.

In the implementation of this approach, some STCGs were slow in developing structured operating procedures, with the result that early compilations of needs (see, for example, DOE, 1994b; 1995g) were sometimes less well prepared than desirable; later compilations were substantially improved. Over time, the needs template used by the STCGs was standardized across all sites (in the summer of 1996), which provided a more uniform basis upon which Focus Areas could base their technology selection decisions. Some Focus Areas (e.g., the Mixed Waste Focus Area) used other available resource material (e.g., Mixed Waste Site Treatment Plans developed by EM-30) to establish their own assessment of site-wide and national needs. The planning efforts associated with *Accelerating Cleanup: Paths to Closure* (DOE, 1998a) also generated data relevant to the definition of site technology needs.

OST has funded the STCGs in the past in order to collect and prioritize DOE-EM site needs. Whether or not this form of collecting site needs will persist, some mechanism to solicit technology development needs from users is necessary, because this function (as depicted by Box 6 of Figure 4.1) is necessary to learn the needs of users. If the STCGs are truly useful to the sites, as was the committee's impression at Hanford site, then perhaps STCG activities can be cofunded by the sites, with the degree of cofunding serving as a way to measure OST's effectiveness in servicing the needs of that site.

Identifying Technology Needs by Characterizing Problem Boundaries and Opportunities

Another OST effort at identifying technology needs was to fund projects whose purposes were to define problem boundaries and technology opportunities. These activities include the Mixed Waste Focus Area's Waste Form Initiative, Thermal and Nonthermal Mixed Waste Treatment Studies, and the HTI. These activities, described below, have not been evaluated in detail by the committee, but are mentioned for completeness to exhibit another way for OST to identify technology needs other than by soliciting requests from the problem owners and technology users of other EM offices.

The Waste Form Initiative of 1997-1998 was a scoping study of the use of existing waste forms and treatment options for the inventory of DOE-EM's mixed waste. The purpose of the study is to determine whether any new waste forms (and associated treatment processes) should be developed in order to properly dispose of the existing inventory. The tentative conclusion of the study is that less than 100 cubic meters—a small fraction of the total inventory—lacks a treatment and disposal option, a result that argues against a large-scale effort to develop new waste forms and treatment processes.

The thermal and nonthermal treatment studies were conducted to compare the total life-cycle costs of competing technical approaches to dealing with DOE mixed wastes. One of the conclusions is that nonthermal treatment facilities, although devoid of the pejorative connotations associated with incineration, cost at least $1 billion more in total life-cycle cost than thermal systems.

The 1997 HTI is an effort to gather technical information on two tanks to inform tank closure specifications. The HTI was endorsed by the Hanford STCG and funded directly from headquarters. It is jointly funded by OST and EM-30.

Aggregation and Prioritization of Technology Needs for Each OST Program (Box 7)

This step was accomplished by OST program units such as Focus Areas and Crosscutting Programs, as described in more detail in Chapter 4 and Appendixes C and D. These program units receive needs from the STCGs, reprioritize them at a national level, and compare the resulting priority needs with the technical products of ongoing technology development project work in order to identify the needs that merit further attention. A lack of adequate information or a lack of technical knowledge may permit individual Focus Areas to inappropriately link new needs defined by the STCGs with existing OST technology portfolios. The consequences are either that a need is misunderstood or that a technology's application to site needs is misrepresented.

The needs statements that the STCGs create contain technical specifications to capture or represent site needs at a reasonable level of specificity. The Focus Areas and other OST program units combine several of these needs statements into work packages, each with one or more technology projects contained within it. This "roll-up" process can be advantageous if one technology can be developed to be flexible enough to meet more than one need. However, this process of representing the program by a suite of work packages can be disadvantageous if important information is lost in the conversion of the site need into more general representative language. That is, the method has a potential weakness if a technology was developed and designed to meet a need that is generally stated, but would have limited applicability to the actual specific circumstances of the individual site needs on which the more general statement was based. OST faces these and other RD&D-related challenges. The solution to this problem is close interaction with the ultimate users, in a way that is reflected in decision making.

Chapter 4 describes the process used by OST to collect and prioritize needs when the problem owner is the community of users within EM-30, EM-40, or EM-60 offices at DOE-EM sites and a Focus Area is the provider program, as shown in Figure 4.1. This figure can also be used to show the user-provider dual program office interaction that occurs within OST for "technology transfer" at more basic phases of RD&D. Focus Areas and Crosscutting Programs can be user programs, receiving technology development projects begun by separately funded OST programs such as the Industry Program, the University Program, the TISA-International Program, and the DOE-EM EMSP. This last program performs mission-directed science, for which the "needs" are appropriately stated at a much more general level.

Table 5.1 depicts the relationships among EM programs, and Appendix E summarizes the means by which needs are identified and prioritized and projects developed in these other OST program units. Greater transparency of the needs identification and prioritization processes in these programs is desirable, and greater uniformity across FAs is needed.

Funding Decisions (Box 8), Soliciting RD&D Proposals (Box 9), and Performing Technology Development Projects (Box 10)

These steps are discussed in Appendixes C-E for each OST Focus Area and Crosscutting Program and for the other OST programs that fund project work, namely the Industry Program, the TISA International Program, and the ASTD program.

TABLE 5.1   Program Office Combinations That Relate as User-Provider Pairs Shown in Figure 4.1

User	Provider
Problem-holding site manager in Em-30, 40, or 60	Focus Area
Focus Area	Crosscutting Program Industry Program University Program TISA International Program EMSP
Crosscutting Program	Industry Program University Program TISA International Program EMSP

Pragmatic Aspects of Funding Decisions Based on Prioritized Needs (Box 8)

Although funding high-priority needs until resources are exhausted is highly desirable, there are factors that complicate this straightforward approach. Three of these complicating factors are described below. They support the general truth that allocation of resources to the highest-priority needs may not lead to optimum results (see, for example, Levary and Seitz, 1990). First, because most projects are multiyear, the technology provider program unit will be considering funding a mix of ongoing projects to meet previous needs and new needs. Thus, the decision-making process must include provisions to determine the relative value of the ongoing projects, based on the results of project monitoring and evaluation, and compare this to the value of meeting a new need. Termination of an ongoing project usually results in the total loss of the investment to date, and for this reason, ongoing projects are typically assigned a high priority and continued unless there is good reason for termination. Valid reasons for terminating an ongoing project include major revision in the importance of the underlying need, a review that indicates a project is unlikely to produce results that will eventually lead to deployment, and severe budget reductions that require terminating lower-priority ongoing projects to fund new, higher-priority needs.

A second complicating factor is that distribution of funding to needs in priority order requires each aggregated need to have an associated budget. For ongoing projects, this budget normally exists. However, for new needs the technology provider program unit must typically estimate a budget in anticipation of specific proposals. Although this frequently requires subsequent adjustments, the problems are generally not too severe because multiyear projects usually begin with a relatively small rate of expenditure, which allows adjustments to be made in future years based on fairly reliable budget estimates.

The third complicating factor is the federal practice of essentially continuous review and revision of budgets and budget decisions. There is a legitimate need for higher-level management to review the performance of the technology provider program unit and ensure that its activities conform to policy goals. However, this has evolved into directives concerning what should or should not be funded at a level of specificity that constitutes micromanagement (PCAST, 1997), coupled with budgetary issues such as withholding promised budgets until late in the year or midyear budget adjustments (usually rescissions). These result in increased program overhead and decreased efficiency of RD&D projects that are abruptly started, suspended, terminated, or redirected. The intervention of headquarters management in program decision making should be reserved for policy issues in general. Further exacerbating this is recent congressional pressure to reduce funding carried over into future fiscal years. Although there are apparently some good reasons for this restriction, the practice can be detrimental to the conduct of multiyear projects, especially those that are funded late in a fiscal year, when a federal budget does not exist early in a new fiscal year or when a project must make major commitments early in a fiscal year. The committee believes that congressional concern could be largely addressed while unburdening ongoing projects by restricting carryover at the program level, but not at the project level.

Pragmatic Aspects of Proposal Solicitation and Provider Selection (Boxes 9 and 10)

The committee believes that the solicitation of technology development proposals should be open and competitive to the extent that multiple proposals are realistically possible. At one end of the spectrum, science needs should be met by proposals obtained through a widely advertised, open solicitation, as is already being done in the EMSP. At the other end of the spectrum, demonstration projects are often tied so closely to specific site applications (e.g., tanks and waste burial sites) and processes that have been winnowed through the multistage RD&D process to a point at which only one choice is possible. In this case, an open solicitation is misleading and wasteful, and should be avoided. Intermediate situations may call for intermediate solutions.

In evaluations such as those described in Chapter 4 and Appendixes B-E, in which a small team of individuals apply their collective judgment, the experience and knowledge of team members is important to the quality of their decision. Therefore, an effort to include individuals with this experience and knowledge base is a way to improve these processes.

Project Monitoring and Evaluation (Box 11)

OST has implemented (DOE, 1996o) a stage-and-gate system (Paladino and Fox, undated; Paladino and Longworth, 1995) to track technology development projects through various stages of development (see Figure 4.2). The purpose of this "stage-and-gate" tracking tool is to provide a disciplined process for assessing and managing the performance of evolving environmental cleanup projects. A project may go through as many as all seven of the stages listed below:

Stage 1: basic research
Stage 2: applied research
Stage 3: exploratory development
Stage 4: advanced development
Stage 5: engineering development
Stage 6: demonstration or pilot operations
Stage 7: full-scale implementation

Each stage has goals, objectives, and measures of effectiveness. For example, in stage 1 the goal is to generate new ideas, an objective is to identify new environmental technology, and measures of effectiveness include technology end user need, technical merit, costs, and safety.

There are gates between adjacent stages. A project at stage x must pass through gate x before it can go to stage $x + 1$. There are three major actions that may be taken at a gate: (1) go forward, (2) hold for specific action, or (3) stop. Gates 2 and 4 are major decision points for OST technology development projects, because later-stage work is typically more expensive to fund than work at earlier stages. At these gates, the Focus Area or Crosscutting Program leadership convenes a review panel to rate (grade and score) specific technologies based on the requirements and criteria established for that gate. The resulting evaluation is forwarded to a review group for concurrence before the project can proceed to the next stage.

The committee suggests the following four ways in which the stage-and-gate project managing and tracking process might be clarified and improved.

Suggestion 1. There should be an explicit "entry" or "validation" decision to confirm that a project is sufficiently meritorious to warrant development and tracking, during project solicitation and selection. This is important because once a project is in the system, it has a high probability of advancing through the process, at least to gate 4.

Although the stage-and-gate process is part of a structured decision-making approach, it should be viewed primarily as a tracking tool. It does not provide information about the initial decision as to what technologies should enter the process for tracking during development. Awareness of the spectrum of technologies available, both external and internal to DOE, should be a requirement for a "make-or-buy" decision on whether a technology need statement should be converted into a technology development project. There did not appear to the committee to be a formal approach in place at all times (see examples in Appendix C) to review existing technologies or technologies already under development and to determine if there was value in developing similar and/or improved technologies.

The Defense Nuclear Facilities Safety Board (DNFSB, 1998), similarly recognized that an increased awareness of technology state of the art and state of practice would improve these decisions. Tools such as existing available technology listings[5] developed and demonstrated by industry and other countries (e.g., the Organization for Economic Cooperation and Development) can be helpful, although they are limited by the specific nature of the technology needs and by timely awareness of them by the potential user. The decision as to what projects are selected is aided by input from a motivated, skilled work force, as exemplified by the Tanks Focus Area's technical team evaluations and the Mixed Waste Focus Area's waste type manager teams.

Suggestion 2. The rating and scoring systems used at gates 2 and 4 have flaws that should be corrected. The pilot reviews (DOE, 1996o) at gate 4 done on two projects used criteria grouped into the following six broad categories:

1. technology need (170 points);
2. technology merit (170 points);
3. cost (190 points);
4. safety, environmental protection, and risk (150 points);
5. stakeholder issues (160 points); and
6. commercial viability (160 points).

The numbers in parentheses represent the maximum number of points out of 1,000 that a project can obtain in the given category. Reviewers of a technology project provide ratings on three to five relevant criteria within each category to derive an (unweighted) average rating. The score for the category equals this average rating multiplied by the maximum number of points for the category. The score for the project is the sum of its category scores.

---

[5]Two Focus Areas—the Subsurface Contaminants Focus Area and the Decontamination and Decommissioning Focus Area—have tasked contractors to develop such databases.

This scoring process seems both unnecessarily complicated and flawed in concept. For example, a project could have scores of 0 in both the need and the safety categories and still receive 680 points. One of the two projects in the pilot review had a score of 654 out of 1,000 and received a go recommendation. The second project had a score of 528 out of 1,000 and was put on hold. The "go" project received a rating of 2 of 10 on the criterion of "end user performance requirements incorporated and implementation issues defined." This low rating deducted only 28 points from the possible score of 1,000, yet it reveals a significant weakness of the project with respect to OST's major difficulty—deployment by the end user.

These examples demonstrate that if the structure or model used for the quantification procedure is incomplete or inaccurate, the results can be misleading or in error. More emphasis must be placed on low ratings in important categories. Specifically, unless a project receives at least specified minimum scores in key categories (i.e., threshold criteria), it should not be given further consideration.

Suggestion 3. Gate 6 and stage 7 should be given much greater attention. OST controls all of the stages and gates except these. However, the overall goal (and implied performance measure for OST) is for its projects to reach stage 7 and be deployed. Perhaps one benefit from the formalization of a technology decision process is that it shows OST's need to exercise influence on gate 6 and stage 7.

Suggestion 4. OST's implementation of a stage-and-gate tracking tool should be simplified to the extent possible by reducing the number of gates at which projects are formally evaluated by personnel outside a particular OST program unit. For example, although the stage-and-gate model has two research stages, only one review might be needed at a project's research stage. Similarly, the three stages and gates at development junctures could be combined into only one review point.

Deployment (Box 12)

· Deployment of OST-developed technology is an important measure of performance and is necessary if OST is to contribute to the technical capability applied to DOE-EM cleanups. However, in the current DOE-EM organization, deployments are conducted by the other EM offices with cleanup responsibilities, as Figure 4.1 illustrates. This technology "hand off" to a different program office makes the deployment step susceptible to programmatic issues (e.g., budget shortfall, change in cleanup priorities or preferred technology, or change in schedule), not just issues of technical viability. Any expanded OST role in deployment diverts its resources from the development of technologies.

Deployment of an OST-developed technology is thus outside OST's direct control. These decisions are made by site managers at DOE sites within other EM offices who own the remediation or waste management problems on which the developed technology is intended to be applied.

It is evident in retrospect that having representatives of other EM offices participating in OST programs has not been sufficient to result in an adequate level of deployment. However, progress has been made with the establishment of the Focus Areas and STCGs, and still more progress in stimulating deployment has been attempted with the recent establishment of the ASTD program and the Technology Acceleration Committee composed of upper-level DOE-EM management. This area could be improved still further by closer, more frequent discussions among Focus Areas, STCGs, and technical staff at the sites.

Deployment can be enhanced with (1) well-defined jurisdictional boundaries between OST and other DOE-EM offices and (2) improvements in the quality and/or frequency of interaction between OST technology developers and DOE-EM personnel knowledgeable about site activities and problems. On the first point, the role of other EM offices should be to provide OST with technology needs and deployment opportunities, while the role of OST should be to develop appropriate technology in view of these user inputs.

SUMMARY AND CONCLUSIONS—SPECIFIC ISSUES

The decision-making process diagrammed in Figure 4.1 is a useful way to describe OST's achievements and interactions with other organizations. The remarks below summarize the box-by-box description of how OST accomplished these process steps in FY 1997-1998.

- The source of funds and the restrictions placed on some funds are beyond OST's direct control and are therefore constraints upon the program.
- Budget allocations to OST program units are done by OST headquarters management.
- Site program planning is outside OST's purview, yet determines technology needs.
- OST has employed STCGs to gather and prioritize these user-specified site needs.
- OST has also funded studies to characterize problem boundaries as a way to identify technology needs.
- Provider programs (such as Focus Areas, in instances where user programs are the site programs funded by other EM offices) are where key decisions are made using two inputs: (1) a prioritized lists of needs and (2) available funds, each of which is subject to changes over time.
- OST's implementation of the stage-and-gate tracking tool can be improved.
- Deployment of an OST-developed technology is outside OST's direct control. These decisions are made by site managers at DOE sites within other EM offices who own the remediation or waste management problems on which the developed technology is intended to be applied.

Chapter 6 presents findings and recommendations based on both the general considerations and the more specific evaluations in this chapter.

6

Findings and Recommendations

The prioritization and decision-making processes considered in this report are those used to fund RD&D in support of the DOE-EM cleanup and waste management missions. This chapter presents the major findings and recommendations. Because the committee believes that there are some important recommendations for effective OST decision making that require actions by levels in the DOE-EM organization above OST, the findings and recommendations are divided into two categories, general and specific. Many of the recommendations on general issues are ones that OST may not be able to address directly; nonetheless, the committee believes them to be important and they are included for possible action by DOE-EM.

GENERAL FINDINGS AND RECOMMENDATIONS

The recommendations that follow are offered to improve RD&D decision making in the current DOE-EM organization,[1] in which OST's purpose is to develop technology to be used by other EM offices, based on their needs. The recommendations below are generally designed to elucidate the technical needs of the user community for OST's use, to improve the connection between these needs and OST development goals and projects, and to enhance OST decision making with modern methods and good practices. OST efforts to implement these recommendations will, in the committee's view, help to address barriers (see Chapter 1) and other current issues (e.g., the right performance measure(s)) for the OST program.

Importance of a Central RD&D Function

Since ends as well as means are important to any decision-making process, brief remarks on these issues are offered. The "ends" are appropriate RD&D activities; the "means" is the OST program office, which in the current DOE-EM organizational structure is (1) separate from the other EM offices with technology needs and users, and (2) centralized in that it is coordinated from DOE headquarters rather than from any EM site or user office. A legitimate issue is how effective a centralized RD&D function such as OST (as opposed to some other organizational structure) can be in funding technology projects that are developed and deployed on DOE-EM site cleanups and whose use provides benefits as measured in cost, risk, schedule, or other relevant criteria.

---

[1]The committee did not address whether other organizational structures were possible to accomplish RD&D in support of the DOE-EM mission, other than to note, as shown below, the potential advantages of a centralized RD&D function for the DOE-EM program.

Finding The DOE-EM program faces many difficult technical challenges. Some pervasive cleanup problems (for example, some of those involving ground water contamination) cannot be solved in a practical sense with current technology. Other problems could benefit by technical solutions that reduce cost and/or risks to the public and the environment. Solutions to problems such as these and others may be provided by developing advanced technologies through a central RD&D organization that addresses more broader-based issues than those that might be addressed by a single site or a single management and integration contractor or its subcontractors. In DOE-EM, OST provides such a centralized RD&D function.

Recommendation to DOE-EM A centralized RD&D function within DOE-EM should be maintained because of its potential advantage in coordinating potentially duplicative technology development efforts needed at DOE-EM sites and because it is in a better position to address important broader issues (e.g., alternative technologies in the baseline functional flowsheets and alternative functional flowsheets) than more specifically directed RD&D.

Deployment of Technologies

Finding OST has experienced a significant problem in increasing the percentage of its technologies deployed at the sites. The problem arose in the past for two reasons. The first is that in its early days, OST failed to attempt to solicit in a structured way technology development needs from problem owners on the site. Consequently, some OST programs were often perceived as not being relevant to site needs. Since it is fruitless to have an RD&D program without deployment of the technologies it develops, OST and EM headquarters have attempted to address this problem by establishing first the Focus Areas and STCGs, then the LSDPs, and in 1997, the ASTD program and the TAC. The second reason is that the sites lack strong incentives to use OST-developed technologies, regardless of agreements or understandings OST may believe it has with the sites. This second reason is outside OST's direct control and is addressed in the recommendation below.

Recommendation to DOE-EM DOE-EM should continue to seek ways to assist OST in getting its developed technologies deployed at the sites. The ASTD program, which serves a useful function of facilitating the deployment of already demonstrated technologies at DOE-EM sites, is not directed specifically toward the technologies developed by OST. Thus, there is still a need for more aggressive action by DOE-EM in deploying OST-developed technologies, in addition to those available from industry.

Recommendation to OST OST's role should be to develop technologies that address DOE-EM site needs. To engender deployments, OST should take the steps listed below to identify current and future DOE-EM site technology needs and should use this information as a guide for tailoring an RD&D program to meet these needs. Use of financial incentives such as those provided to the sites by the ASTD program are not desirable but appear to be necessary at the present time. They should be discontinued as soon as practicable because they commit OST funding that could be redirected with advantage to other OST initiatives. For identification of current needs, OST should continue to stress the involvement of the sites in OST technology development and procurement activities and, reciprocally, should involve itself to the extent possible in the site activities and problems.[2] For identification of needs likely to persist in the long term, OST should identify technology vulnerabilities (i.e., process steps with significant probability of technical failure) in baseline functional flowsheets and use this information to develop backup technologies and alternatives that have been agreed on through discussion with the site problem owners.

Balancing Research with Development and Demonstration

To have a balanced RD&D program, OST should support research-oriented activities that explore novel and innovative ideas, as well as more development-oriented activities directed at existing processes for specific applications. Research-oriented projects involve relatively high technical risk and are much less than 100 percent successful, if success is measured by large-scale application (i.e., deployment).

Finding A balanced RD&D program will have a less than 100 percent success rate, if the measure of success is deployment in the relatively near term.

Recommendation to DOE-EM The percentage of OST technologies that reach the deployment stage should not be the sole figure of merit used in judging the OST program, although it is an important one. A long-term view should be adopted wherein the direct use of OST technologies or the use of derived technologies is also considered in the evaluation of OST's portfolio of technology development projects.

Site Baseline Remediation Functional Flowsheets

Finding Decisions on site technology development needs are derived from the so-called baseline functional flowsheets developed at the sites by the site remediation problem owners. These flowsheets are fundamentally important in determining process steps and related technology needs. In effect they dictate the technology needs to which OST must respond with technology development programs. As a consequence of the current DOE-EM organizational structure, OST has no direct role in establishing the baseline functional flowsheets, which are developed by other EM organizations and contractors at the site level. At present, to influence decisions on technologies to be adopted, OST must undertake studies that compare existing baseline technology costs with more favorable costs of OST-proposed technologies and in this way try to persuade site managers to adopt technologies different from those they are already committed[3] to use.

Recommendation to DOE-EM The expertise of technology developers supported by OST could be of value to the site problem owners in formulating and maintaining technically sound and practicable cleanup functional flowsheets. Therefore, efforts should be made to have substantial involvement of appropriate OST and OST contractor personnel in reviews of functional flowsheets. Such participation

---

[2]One practice in the past to enhance this involvement has been for the site user representatives of the Tanks Focus Area to endorse the consolidated, prioritized list of technology needs (see Appendix C).

[3]If these commitments are represented in legal agreements or in site contracts, opportunities to make decisions on alternative technology may be significantly restricted.

would have the benefits of (1) ensuring that OST technology developers fully understand the site problem owners' technical needs and their bases and (2) increasing the sites' confidence in OST's dedication and ability to meet their needs.

Technical Alternatives to Baseline Remediation Functional Flowsheets

Finding Good decision-making practices (e.g., hedging against technical uncertainty and insisting on alternatives) imply that DOE-EM should plan for alternatives to the site baseline functional flowsheets, especially when the baseline flowsheet involves high cost, high or poorly defined risk, and/or substantial probability of technical failure. Uncertainties about the future, both in funding and those related to possible failures of existing functional flowsheet process steps, highlight the importance of developing technology alternatives. In addition, the possibility of failure of large-scale privatization contracts to achieve site cleanup goals reinforces the value of technology developments targeted at alternative functional flowsheets. At present, there is little or no DOE-EM funding for developing alternative functional flowsheets, and consequently, no OST activity is expended on technology development directed toward alternative functional flowsheet technology needs.

Recommendation to DOE-EM The development of alternative functional flowsheets is the responsibility of DOE-EM offices other than OST, but they should seek OST's input. It is highly desirable that the problem-owning EM offices should seek out and acknowledge the potential vulnerabilities—in cost, risk, and technological failure—of the baseline functional flowsheets and processes and, with OST's assistance, develop alternative flowsheets as appropriate. OST should encourage this course of action and seek to collaborate in it.

Recommendation to OST OST should attempt to make input to alternative functional flowsheets and, in particular, should advocate their development when the baseline functional flowsheet involves high cost, high or poorly defined risk, and/or substantial probability of technical failure. OST should identify specific technology development opportunities aimed at supporting alternative functional flowsheets and processes designed to enhance the overall probability of remediation successes and to minimize program delays. In practice, this means that OST should be allowed to commit a portion of its resources to developing technologies to address needs derived from such alternative functional flowsheets, in addition to developing technologies to meet the needs derived from the baseline flowsheets.

Independent, External Reviews

Finding Peer reviews—that is, reviews by technical experts who are independent of and external to the program of work being reviewed—are a vital part of a credible decision-making process. OST programs and projects are subjected to many, often resource-intensive reviews but, according to recent NRC reports (1997b, 1998b), to few true peer reviews. To address this shortcoming, in August 1996 OST established a peer review system using the American Society of Mechanical Engineers (ASME, 1997a) to review works in progress. The result is an OST program with many different reviews, which are not fully effective in the important decisions of the initial selection and subsequent continuation of meritorious technology development projects.

Reviews of Technology Development Projects

Recommendation to OST Peer reviews of technology development projects should be part of OST's decision-making process. These project peer reviews should occur as necessary and in a way that is not an overly burdensome commitment of OST resources. Therefore, the OST review system should be streamlined by reducing the number and types of reviews based on an analysis of the objectives being served by each. Reduction in the number of reviews could be accomplished in part by combining reviews where practicable.

Reviews of Program Budget Targets and Their Rationales

Review of technology development projects is part of RD&D project monitoring and evaluation, which is but one step in the OST decision-making process (see Figure 4.1). The committee therefore considered whether other steps of the process merited review.

Finding The formulation of budget targets for the various OST program units (e.g., Focus Areas and Crosscutting Programs) is an important decision point in OST's process. At this decision point an independent external review is needed but is not now being undertaken. These funding targets are made by the OST upper management levels at headquarters based on input from DOE site managers and other OST program managers. These targets are proposed to higher authorities (and ultimately to the Congress) through the annual federal budget process.

Recommendation to OST An independent, external review should be held on the basis of, and rationale for, decisions on funding targets within OST. One goal of this review should be to identify the technical areas of greatest need, where improvements over existing conventional approaches would have the greatest benefit to DOE-EM. This review and its outcome should take into consideration such factors as DOE-EM programmatic strategies, political pressures, stakeholder pressures, risk to human health and the environment, safety, cost-benefit, and timing. Such a review might be carried out by an already-constituted authoritative external body such as the Environmental Management Advisory Board (EMAB), or by a group created specifically to conduct the review.

OST Involvement in Reviews of Site Remediation Functional Flowsheets

Finding The site remediation functional flowsheets, whose development and review are in the domain of the site problem owners, are important for OST because they define the user plans from which technology needs are derived.

Recommendation to DOE-EM Site remediation functional flowsheets should be subjected to independent, external review before they are adopted, and periodically during development of the technologies that are to implement them. The EM offices developing these flowsheets should have them peer reviewed, that is, reviewed by technical experts who are independent of and external to the program. This expertise may be found in academia, private industry, and national laboratories. The purpose of such reviews is to identify possible vulnerabilities or uncertainties in the functional flowsheet assumptions and technology selections. The committee understands that the other EM offices already sponsor such peer reviews of functional flowsheets for the most part, but it would recommend the practice for all important functional flowsheets.

Recommendation to OST OST should try to work in conjunction with other DOE-EM offices to participate to the extent possible (e.g., by establishing a role for OST contractors) in the schedule of peer reviews of the site baseline functional flowsheets.

The purpose of this OST involvement is to make the results of these reviews and the rationale for the technology development needs available to OST. This OST role in reviews of site baseline functional flowsheets will also, in the committee's view, help strengthen the interaction between OST and the DOE-EM user community.

SPECIFIC FINDINGS AND RECOMMENDATIONS

The recommendations presented below are within OST's purview and control to implement. This is not to say, however, that exigencies of the moment regarding such constraints as limited funding may not prevent OST from taking immediate action on some of them.

OST's Institutional Environment Affects Customer Interactions, Relevance to Site Needs, and Deployment

Finding OST operates in a complex, ever-changing, and politicized environment. An important complication for OST is that the sites and the EM offices that are responsible for cleanup activities have a great deal of autonomy in selecting baseline remediation processes and technologies to deploy. Furthermore, they are under no obligation to use OST-developed technologies, even if they have made earlier commitments to do so. As new cleanup and waste management problems are found at the sites, new technology needs arise. This creates a very difficult situation for OST in keeping abreast of technology needs and getting its technologies accepted and deployed.

Recommendation To the extent possible, OST should increase its efforts to identify site technology needs on a current basis and to anticipate future needs. Regularly scheduled meetings with site problem owners should be considered. More discussions of technical issues and their implications for technology development needs should be held with the working-level scientists and engineers.

Finding OST has more than one "customer" to satisfy. Other EM offices such as EM-30 and EM-40 are obvious end users of OST technology developments, but the U.S. Congress must also be satisfied that a reasonable fraction of OST products are useful and worth their cost. Furthermore, parts of OST expenditures are congressionally mandated (U.S. Congress, 1994; 1995a-b). Consequently, the type and quality of the information provided to Congress (and to other interested review groups) are critically important to OST.

Recommendation OST should ensure that the decisions underlying the technologies it develops are well documented, traceable to customer needs, and supported by sound technical reviews. Records should be kept of the reasoning by which the deciding factors were evaluated, including whatever method(s) were used in their evaluation.

Finding Although the user focus is necessary, not all RD&D needs come from user requests. As an example, alternative technical approaches to site remediation baseline functional flowsheets are a potential source of technology development needs, and these alternatives would not be derived from user requests based on baseline functional flowsheets. Therefore, there is a need for exploratory RD&D with a shorter time frame than the longer-range basic research projects of the Environmental Management Science Program, to meet the need of alternatives to baseline functional flowsheets (see Chapters 3 and 5 and Appendix E).

Recommendation Although the technology development projects should be based primarily on specific needs at the sites, some should be of an exploratory nature to meet the need for backups and alternatives to the baseline functional flowsheets.

Top-Level Strategic Planning and Goals

Finding OST's strategic goals do not provide an adequate level of guidance for program managers as they attempt, in collaboration with users, to assign priorities to technology needs. In times of limited funding, it is especially important for managers to have the type of guidance and direction that come only from strategies developed by top OST managers, whose perspective is much broader than that of personnel further down the management chain.

Recommendation OST managers, in conjunction with other top-level EM managers,[4] should

- produce strategic goals and plans that define explicitly the technical problems that the program will (and will not) address, and
- use these goals and plans effectively within OST program units to assist them in making technology development decisions.

Although EM "strategic goals" are now available (DOE, 1997j), these are obtained through a user-driven and therefore "bottom-up" decision-making process of other EM offices in which most remediation decisions are made at individual sites (DOE, 1998a; 1998b). Therefore, these strategic goals,

---

[4]Input from other EM offices is recommended since OST goals should be derived from user plans and needs.

which target the sum of the work to be achieved at all sites in a bottom-up approach, differ from the top-level goals recommended here for OST that are specified by a centralized, coordinated, top-down process to drive the technology development program. Of course, any top-level strategic goals developed by OST should be consistent with the EM mission and be derived in concert with user plans and needs. That is, any such strategic goals for OST would be guided and constrained by priorities established within other DOE-EM offices, priorities that have varied over the years and would be represented by their top-level goals and other statements of their high-priority tasks. A particularly important area for current attention relates to the implications of privatization for OST's technology development program.

This goal-oriented approach, to be successful, would have a significant impact on the way in which OST operates. Obtaining the "right" goals is a challenge, as is the orientation of program activities in support of these goals. For the goals to work, everyone in the organization that can have an impact on goal achievement must be measured on the success of reaching the goal (or at least the part of the goal that their job impacts). If these broadly written goals are in place and directions of important efforts can be derived from them, then this framework would provide a way to readily assess the value of each RD&D project.

The committee chooses not to be prescriptive, but instead believes that OST management should be responsible for formulating its goals and criteria. These would include (1) adequate cleanup targets (e.g., level of decontamination to be achieved in various media, for use as program goals by Focus Areas and other OST program units in soliciting and selecting projects) and (2) a set of objective criteria to be used in evaluating technology development proposals and projects. This effort could solicit and use inputs from the users of technology (i.e., site contractors) and all interested parties.

Use of a Structured Decision-Making Process

The main reasons for a structured decision-making process are to ensure its comprehensiveness, objectivity, and consistency. Often, these qualities can best be realized when data and other information are expressed in quantifiable terms. Quantifiable inputs include not only measurements that are already in quantitative form (e.g., contaminant concentrations or dollar values), but also inputs such as expert judgment estimations (e.g., scores expressed as high, medium, or low) that can be scaled and converted into quantitative form. Both quantitative input (such as cost estimates) and expert estimates involve judgment; the aim of a quantifiable approach is to make these judgments as objective and transparent as possible.

Finding Structure to the OST decision process is warranted, in part because of the complexity of OST's institutional environment and in part because of the potential for change over time in technology needs (which serve as the basis for the technology development projects that OST funds) and top-level priorities.

Recommendation For decisions involving the allocation of significant resources, OST should institute a decision-making structure wherein projects and/or proposals are evaluated against consistently defined criteria such as project cost, probability of technical success, probability of implementation on field applications, potential cost savings, and human health risk reduction. This structure should be applied broadly throughout the organization, with each OST program unit evaluating projects against the same criteria. An important criterion would be the project's relevance to site activities; that is, all projects should have one or more specific objectives related to the cleanup and waste management goals at sites. For an ongoing OST project, its history of past performance in meeting appropriate developmental objectives would be useful information to gauge its likely future success.

The quantities mentioned in the preceding paragraph are recommended to be estimated in order to provide more discipline and guidance to decision making than if these estimates were not done. It is a challenge to perform estimates well with the degree of rigor that is possible depending on the state of maturity of the concept. Initial estimates of early-stage development projects would likely be subject to data limitations, large uncertainties, and limited sophistication. Refinements to these calculations would be likely over time as a technology development project progresses and should be documented. The quantification of project benefits and probabilities of success is necessarily uncertain, but the committee believes that a rigorous attempt to estimate and use these quantities and other available data is preferable to the less structured methods that OST has used in the past. The recognition and evaluation of uncertainties is itself a key part of a decision process.

This structured system is proposed as a useful decision support methodology to organize inputs that are used by the OST managers who make decisions. Any such decision support tool has limitations, particularly if it fails to account for all significant factors; [5] therefore, managerial flexibility to "override the numbers" in occasional situations is necessary. There should also be sufficient managerial flexibility to enable responses to late-breaking development opportunities.

Industrial RD&D Decision-Making Practices Applicable to OST

Finding Good private-sector decision-making practices include the use of formal decision-making processes that employ quantifiable, measurable goals and follow-up procedures such as those discussed in Chapter 3 for RD&D decisions.

Recommendation OST should adopt, where applicable and appropriate in the OST environment and to the extent practicable, the basic principles of private-sector formal decision-making and follow-up practices that are presented in Chapter 3. In particular, an attempt should be made to assess the following factors and adopt them consistently where applicable across the entire organization:
- Understand, focus on, and monitor changes in customer needs and requirements.
- Agree on clear and measurable goals.
- Use a formal (i.e., common, consistent, structured, and rational) technology development decision-making process and apply it uniformly.
- Think strategically (i.e., long-term and high impact).
- Measure and evaluate to guide resource allocation.

---

[5]Indeed, because of the potential for complexity in the various considerations relevant to a decision on technology RD&D, the phenomenon of "rank reversal" (Corbin and Marley, 1974; Buede and Maxwell, 1995; Saaty, and Vargas, 1993; Tversky and Simonson, 1993) is possible, resulting in a decision in which the outcome that is preferred is not the one that scores most favorably against all the criteria used in the ranking or rating exercises. The committee suggests that managerial flexibility be invoked to provide for this kind of situation (i.e., to make final decisions by choosing the options that do not necessarily have the highest scores as evaluated against the criteria used) in the absence of a more rigorous method.

- Communicate across organizational boundaries (i.e., with technology users).
- Continually improve the research and development (R&D) management process.
- Hire the best people possible and maintain expertise.

The committee prefers to leave to OST management the task of translating how these practices could best be adapted. For example, "hire the best people possible and maintain expertise" would refer in private industry to the cadre of researchers within the company; for OST, which is a collection of program units run by program managers who contract out the technology development work to PIs, the application of this concept might be best translated as "select the most competent contractors."

The general recommendation is that OST seek to implement those practices that have been identified as improving decision quality. The support for this recommendation comes from published literature on broad surveys of company practices. To lend support to the validity of these decision-making practices for OST, the committee sought confirming evidence that they were useful concepts in organizations analogous to OST in their structure, institutional environment, and/or the kinds of technology within their environmental remediation RD&D programs. The industries cited in Chapter 3 and Appendixes G-I provide examples of institutions that have some similarity to OST. The institutions described there do not represent an exhaustive list of such analogous institutions, but were chosen as specific examples to illustrate the utility and practical application of the good decision-making practices and principles described in Chapter 3. These institutions should not be construed as constituting a representative sample of all institutions employing successful decision-making practices; many such institutions exist, some employing different (e.g., non-utility theory based) methodologies.

Specific Methodologies

Finding The quantifiable approaches used by OST (see Appendixes B-E) are simple multicriteria scoring techniques in which criteria are developed with relative weights assigned to them. A project or proposal is rated against these criteria by reviewers who assign numerical scores. The scores and relative weights are used to calculate an overall score, which is used for planning and management purposes in comparing various projects or proposals.

This simple scoring technique is one example of multi-attribute utility analysis. Utility analysis has its strengths but also its weaknesses. For example, utility analyses can be criticized on several accounts, such as

1. the inability to formulate a meaningfully weighted utility function that adequately synthesizes many, often disparate, "customer" views (Zeleny, 1982);
2. the possible interdependence (in practice) of criteria that, in utility analysis treatments, are treated as independent (Machina in Stigum and Wenstop, 1983);
3. the problem of expressing preferences for these criteria in ways that cannot be adequately captured by numerical scales of numbers that are added or multiplied together (e.g., translation of measurement scale values, such as those of ratio scales, into interval scale representations) (Saaty and Vargas, 1993);
4. the combination of tangible and intangible factors (an example of the latter is the valuations applied to expert judgments), especially in hierarchical prioritizations (Saaty and Vargas, 1998); and

5. the possible violations of the "principle of invariance," also known as regularity, the independence from irrelevant alternatives, and "rank preservation," by which is meant that the rank order of selections may change as new criteria are added or old ones deleted (Tversky and Simonson, 1993; Tversky, Slovic, and Kahneman, 1990).

Other decision-analytic methods, such as the analytic hierarchy process, have been advanced as alternatives to utility analysis that treat these complicating issues in a different way (see, for example, Saaty, 1994; 1996). The issue for OST applications is which of all possible decision methodologies is adequate and convenient to use for making the requisite decisions. One way to answer this question is to judge whether a method's shortcomings, such as those enumerated above, are important "fatal flaws" that prevent the best decision from being made. The committee makes no finding on a relative comparison of various decision methodologies, or which is "best" for OST, but would instead offer the following recommendation:

Recommendation OST should examine the efficacy of the sets of criteria and scoring techniques currently used by OST program units (e.g., STCGs, Focus Areas, and Crosscutting Programs). This could be done by (1) using one or more contractors with suitable expertise to survey alternative decision-making analytical methods and (2) using the considered judgment of OST management to identify those analytical methods that are well-suited to OST's various needs. The judgments for OST management are whether, for OST decisions, the shortcomings of any decision methodology are important practical concerns and whether the decision methodology provides a useful organization of input information to help guide a decision among several candidates. The current OST quantitative schemes are multi-attribute systems with criteria and relative weights, and rely on expert (e.g., reviewer) judgment to score projects and project proposals against the criteria. In practice, the recommendation offered above would evaluate whether this process is sufficiently adequate or whether an alternative process would provide a more useful output. For example, one way to minimize "rank preservation" problems (issue 5 above) is with the use of pairwise comparisons of alternative proposals (i.e., rank ordering them by having reviewers compare them using their judgment, experience, and knowledge) rather than rating each proposal against a set of criteria and comparing the numerical outputs of the rating and scoring exercise.

One such comparison of multi-attribute value theory and the analytical hierarchy process methodologies showed that the decision outcome depended to a lesser extent on which methodology was used and to a greater extent on issues associated with the way the problem was structured and valuations and weighting factors were elicited (Buede and Maxwell, 1995). As noted above, the general recommendation of this report is for a method with structure, documentation, and quantifiable attributes, without a recorded preference on which specific method with these attributes OST should adopt. The recommendation is primarily that decisions be structured, by which is meant that the goals, factors, and criteria believed to influence the decision should be clearly specified. Records should be kept of the reasoning by which the deciding factors were evaluated, including whatever method(s) were used in their evaluation.

Most decisions on the selection and continuation of any research and development project are revisited several times (i.e., at least annually), particularly if the project is funded during more than one fiscal year. Record keeping would support systematic refinement of decisions as more information is developed and would provide the basis for "lessons learned" to help improve decision quality.

Summary

A major goal of this report is to outline the institutional environment in which OST operates, as the "macrolevel" context within which OST prioritization and decision making occur. To suit this context, a decision method is proposed in a previous section of this chapter that offers structure, quantifiable attributes, and documentation. A "microlevel" recommendation is not offered on which of any of the specific methods available in the literature to meet these general criteria is the best for OST to select and adopt. Rather, a familiarity with various decision-making approaches, accompanied by the considered judgment of OST management, is recommended to direct the selection and adaptation of specific decision-making methods.

Project Selection and Evaluation Criteria

Finding There are no general, OST-wide guidelines for setting criteria for the selection and prioritization of technology development needs, although individual OST program units have developed their own guidelines. Moreover, the general criterion that technologies should be applicable to multiple sites, while useful, is flawed when applied without exception because it may lead to the failure to develop technologies for potentially very important problems that exist at only one or two sites.

Recommendation To the extent practicable, and with input from its various organizational elements, OST headquarters should establish general selection and prioritization criteria, and guidelines for applying these criteria, to include allowance for instances in which exceptions to the criteria may be appropriate.

Procurement of Externally Demonstrated Technologies

Finding The present approach to technology procurement wherein both OST's Industry Program and other OST organizational units such as the LSDPs perform some aspects of technology selection and procurement from industry is cumbersome and duplicative, and impairs OST's deployment initiatives.

Recommendation A better-coordinated, less duplicative, and less cumbersome system should be established for integration of technology procurement activities. Since procurements should be made only if warranted in a make-or-buy decision, the ability to assess available technology is crucial. These assessments should be done, by the use of a comprehensive database of demonstrated and commercially available technologies.

The purpose of these comments is to engender a coordinated program in which information flow is effective and cumbersome or duplicative elements are reduced.

Project Monitoring

Finding OST monitors technology development projects in several different ways, depending on the specific project. One established formal process that OST has adapted and used is the stage-and-gate system. In this system, funded projects are reviewed as they progress through their various stages of development from early exploratory studies to engineering development and finally to deployment. At the end of each stage there is a gate, which is in actuality a decision point on whether to proceed to the next level of technology development. However, the stage-and-gate system does not help to critically review the initial decisions to select projects for funding and entry into the tracking system. In addition, the system has an excessive number of stages and gates to use them all for review points in the OST application, and the scoring system has some flaws.

Recommendation OST should use the minimum number of stages needed to track projects. This will reduce the administrative load and will lead to better decisions by producing better-defined decision points and clearer lines of demarcation between them.

Recommendation In selecting a new technology development project for funding, OST should base this decision on both technical merit and quantifiable estimations of the project's probable value to site cleanup activities. OST has developed this latter concept as part of the criteria of the stage-and-gate system, but OST program units do not uniformly adopt and use these criteria to guide their selection of new projects for funding.

Recommendation OST should correct the additive scoring system to account better for threshold criteria. One way to do this is to multiply scores in key categories rather than add them.

Cost Estimates

Finding OST has attempted to gain acceptance of new and/or improved technologies by showing cost savings. However, it is difficult to assess the claimed savings because they require knowledge of both the cost of the baseline technology to be replaced and the cost of the OST technology. Often the baseline technology cost estimates, if they exist at all, are uncertain. The same thing is true of the proposed replacement technologies. Nonetheless, these cost comparisons can be an important decision-making criterion and can serve a useful purpose if they can be trusted. Uniformity in cost calculations (both information and methodology) is a need that has recently been acknowledged by the Decontamination and Decommissioning Focus Area through its use of the U.S. Army Corps of Engineers as an independent, external organization to perform cost calculations. However, further steps are needed to establish the credibility of cost estimates for OST technologies.

Recommendation OST should do cost avoidance or ROI calculations on its more expensive technologies in a more uniform and credible manner than in the past and should communicate results to the potential technology users in the most effective way possible. Initial estimates of costs and benefits should be developed at the inception of large RD&D projects, and periodic refinements of the estimates should be a part of the project as it progresses.

Exploratory Development

Part of the decision-making process is the gathering of technology needs. The recommendation above on using site remediation functional flowsheets, including consideration of the baseline version as well as alternatives, to best derive these needs will lead to related technology development opportunities.

Finding A fraction of OST's budget is devoted to exploratory *research* through the EMSP, which is a long-range basic research program that stresses research related to environmental cleanup problems (discussed further in Appendix E). Although this program is important, it is not sufficient to meet the need for exploratory *development* studies related to alternatives to the baseline functional flowsheets. These studies are intermediate in timing between the long-range time frame of EMSP projects and the immediate need for work on existing functional flowsheet technologies.

Recommendation Additional funding should be sought (or some existing funding redirected) for exploratory development that is directed to technologies for alternative functional flowsheets.

Because exploratory development is recommended, the issue is raised of how best to fund and manage such efforts. One relatively straightforward way to fund small-scale, exploratory projects in a manner similar to the ways in which some OST funds (specifically, the research opportunity announcements and the EM Science Program funds; see Appendix E) are dispersed and managed at present is to set aside some uncommitted funds (e.g., a fraction of the total OST budget) in a program that disperses them by having an open solicitation to which prospective PIs can respond, with frequent (e.g., quarterly) reviews of submitted proposals by teams of independent reviewers assembled for this purpose. An approach of this kind can capture some of the key features of exploratory decision making in a mechanism that will work in the federal budget system. The recommendation is for OST to adopt a suitable method that will disperse uncommitted funds in a reasonably timely and well-defined way for exploratory development activities.

CONCLUDING PERSPECTIVE

Consideration of OST's decision-making process begs the issue of what decisions are and should be made, the answer to which requires consideration of the role and function of an RD&D organization such as OST. There is a role for RD&D activities in providing economical, effective, acceptable, and practicable technologies for use in DOE-EM site cleanups. Although OST accounts for a small part of the DOE-EM budget, its work can have substantial and beneficial impacts on reducing the costs, risks, and probability of technical failure associated with environmental remediation activities, which are estimated to represent in excess of $100 billion dollars.

OST is an RD&D office within DOE-EM that is separate from the other EM offices that bear responsibility for addressing cleanup problems. This organizational structure is a reasonable one for conducting RD&D and is mirrored in utility consortia such as the GRI and the EPRI, in private industry, and in the Defense Advanced Research Projects Agency of the Department of Defense. Although the findings and recommendations presented in this report are directed to OST, they could, in principle, be applied to any RD&D office that manages project work in support of a broader federal program, particularly if organizational similarities to OST and DOE-EM exist.

RESPONSES TO ISSUES IN STATEMENT OF TASK

The statement of task for this study asked the NRC study group "to evaluate the effectiveness of the OST decision-making process and make specific recommendations to improve it, if appropriate." To accomplish this task the panel was asked to address four points. The material to address these points has been presented in the preceding chapters. This material was used to provide input to the responses to the four points, which are reproduced below.

1. The appropriateness and effectiveness of decision-making processes currently in use by OST to select, prioritize, and fund RD&D activities, both at sites and at headquarters:

The current OST decision-making practices, which change with the passage of time, are embodied in its program structure. The program structure of OST's core RD&D functions is based primarily on STCGs, Focus Areas and Crosscutting Programs, and a headquarters oversight activity. A summary of the elements of this program structure is presented in Box 2.1.

The process functions reasonably well to prioritize technologies to propose for funding at individual sites and to make decisions for funding among sites within the framework in which OST operates. However, as pointed out in this report, some of the process steps—for example, technology prioritization by the STCGs—are cumbersome and ill-conceived, and there has been little effort to apply a carefully thought-out system uniformly across OST. This leads to difficulties in comparing and evaluating priorities across the diverse technology areas that OST's activities encompass. Also, the stage-and-gate process for evaluating technology developments at various stages of maturity is unnecessarily complex.

More direct input by OST to functional flowsheet review at the site level is desirable. However, a major point made in the report is that key decisions about which technologies to support lie outside OST's direct control. Input to technology and functional flowsheet selection at the sites is one of these key decision areas. Another critical point at which peer review is needed and is presently absent is at headquarters where final funding decisions are made.

2. The technical factors appropriate to consider in the decision-making process for selection, prioritization, and the development of cleanup technologies, and the adequacy with which these factors can be measured:

The committee has identified three technical factors to consider: (1) the degree of maturity of the technology, (2) the ease of integration of the technology into a total system, and (3) the impact of the technology.

The *degree of maturity* of a technology is represented by a developing technology's remaining assumptions and uncertainties that require testing, and can be measured in terms of the estimated time required to bring it to a sufficiently large-scale (e.g., pilot-scale) demonstration. It must be recognized that OST is not constrained to use technologies that are mature and, in fact its, greatest contributions may be in the development of new technologies. This report points out that OST should earmark a part of its funds for exploratory development of new technologies that have easily recognizable applications. The stage-and-gate process is designed to assess the maturity and likelihood of success of a developing technology and to terminate technology developments found to be lacking promise of successful completion or application.

The *ease of integration* of the technology into a total system is a question that cannot be answered solely by considering OST and its decisions. This integration depends strongly on the attitudes and constraints of the sites with respect to what they are willing to accept or what they can accept because of regulatory controls and agreements that may have been made with state or other influential bodies, as has been stressed in this report. These factors often control what technologies can be integrated in cleanup systems and therefore represent valid consideration in RD&D decision making.

The intended *impact* of a technology is to reduce costs, schedules, and/or risks of various kinds, and/or to enable a difficult or intractable task to proceed. These intended beneficial impacts imply tradeoffs; they are not necessarily mutually reinforcing. One way to assess these impacts is to measure the number of functional flowsheets needing that technology, the number of times it is needed in each flowsheet, and the importance of the flowsheet need addressed by the technology. The practice of making quantitative assessments of this kind would lessen the need for OST to rely upon guidelines as a basis for decision making. For example, one of the most important guidelines OST uses to decide on funding technologies is their applicability at more than one site. Although the committee agrees that this is indeed an important consideration, there may be instances when a technology is very important for a single site. The quantification of technology impacts would enable an informed decision to be made in estimating whether a single-site technology had greater potential benefit than a technology that had uses at multiple sites.

3. Recommendations, if appropriate, for improving the decision-making process:

Recommendations are addressed in several ways in this report. Specific recommendations are given above in this chapter, separated into general considerations and those specific to the current OST decision-making process and program structure. Chapter 5 presents general principles derived from the private sector that, if followed, will lead to improved decision making. As stated in Chapter 1, the committee did not undertake to prescribe or recommend one methodology or school of thought in decision making over any other in its study.

4. The role and importance of effective peer reviews in the decision-making process:

The committee recognizes the crucial role played by peer review at appropriate points in the decision-making process, and subscribes to the process and opinions described in earlier NRC (1997b, 1998b) reports. These reports also acknowledges that OST is subjected to many reviews (many of which have not been carried out in the manner recommended by NRC, 1998b) and suggest that the number be reduced by combining reviews whenever possible.

The DOE-EM RD&D decision-making process has two key decision points other than technology development project reviews that are opportunities for reviews. These reviews, although not true peer reviews in the sense of NRC (1997b, 1998b), nevertheless can embody many of the features of peer review. This report recommends (1) a "policy review" at the headquarters level where budget targets are formulated for the various OST program units, and (2) OST participation in (and/or access to) external reviews of the baseline flowsheets developed by the other EM offices.

Cited References

Alm, Alvin. 1996. Accelerated Environmental Cleanup. Office of Environmental Management. Washington, D.C.: United States Department of Energy. (Alm, 1996).

Alm, Alvin. 1997. Technology Deployment Memorandum from Alvin Alm. Washington, D.C.: Department of Energy. July 3. (Alm, 1997).

Alm, Alvin, and Modesto Maidique. 1996. Correspondence, w/attachment regarding X'Change 97. Washington, D.C.: Department of Energy. October 23. (Alm and Maidique, 1996).

American Society of Mechanical Engineers (ASME). 1997. Assessment of Technologies Supported by the U.S. Department of Energy Office of Science and Technology: Results for the Peer Review for Fiscal Year 1997. American Society of Mechanical Engineers. (ASME, 1997a).

American Society of Mechanical Engineers (ASME). 1997. Peer Review of MAG*SEP[SM]. Review on April 23-25, in Atlanta, GA. (ASME, 1997b).

Barainca, Michael. 1998. New OST Process for Prioritization: Using the *Accelerating Cleanup: Paths to Closure* Data to Make Budget Decisions. Paper presented to Committee on April 2. (Barainca, 1998).

Bauer, Carl O. 1996. Implementation of New Peer Review Policy. Washington, D.C.: Department of Energy. Memorandum for Distribution. (Bauer, 1996).

Bedick, Robert C. 1997. Industry and University Programs Overview (Viewgraphs). Presentation to Committee on May 9. (Bedick, 1997).

Beitel, Dr. George A. 1996. Mixed Waste Focus Area Program Management Plan. Idaho Falls, ID: Department of Energy, Idaho National Engineering Laboratory. October. (Beitel, 1996).

Beller, John. 1997. Idaho National Engineering Laboratory Environmental Management Technology Development Needs and Opportunities (Draft). LMITCO, Rev. 0.5. Idaho Falls, ID: Department of Energy. (Beller, 1997).

Berkey, Edgar. 1997. EMAB Perspective on OST Decision Making. Presentation to the Committee, November 10. (Berkey, 1997).

Berkey, Edgar. 1998. Review of Focus Area Program. Paper presented to Department of Energy, Environmental Management Advisory Board on January 21. (Berkey, 1998).

Blush, Stephen M., and Thomas H. Heitman. 1995. Train Wreck Along the River of Money: An Evaluation of the Hanford Cleanup. Report presented to Unites States Senate Committee on Energy and Natural Resources. March. (Blush and Heitman, 1995).

Bournakis, A.D., and G.D. Pine. 1997. Benefits of GRI R&D Results That Have Been Placed in Commercial Use in 1992 through 1996. Report No. GRI 97/0164. Chicago, IL: University of Illinois at Chicago Energy Resources Center. (Bournakis and Pine, 1997).

Buede, Dennis M., and Daniel T. Maxwell. 1995. Rank Disagreement: A Comparision of Multi-Criteria Methodologies. Journal of Multi-Criteria Decision Analysis, Vol. 4: John Wiley & Sons, Ltd. pp. 1-21. (Buede and Maxwell, 1995).

Chernoff, H. and L. Moses. 1959. Elementary Decision Theory. New York: John Wiley & Sons. (Chernoff, and Moses, 1958).

Christy, C. Edward. 1997. Industry Program Implementation. Presentation to Committee on May 9. (Christy, 1997).

Conner, Julie, and Mike Connelly. 1996. Mixed Waste Focus Area Briefing to the National Academy of Sciences on November 12. (Conner and Connelly, 1996).

Cooper, Robert G. 1993. Winning at New Products: Accelerating the Process from Idea to Launch. Reading, MA: Addison-Wesley. (Cooper, 1993).

Corbin, Ruth, and A. A. J. Marley. 1974. Random Utility Models with Equality: An Apparent, but Not Actual, Generalization of Random Utility Models. Journal of Mathematical Psychology, Vol. 11: pp. 274-293. (Corbin and Marley, 1974).

Defense Nuclear Facility Safety Board (DNFSB). 1998. Information from website (www.dnfsb.com.) accessed in October. (DNFSB, 1998).

Department of Energy. undated. INEL Identified Technology Development Needs Prioritization. Washington, D.C.: Department of Energy. (DOE, undated a).

Department of Energy. undated. INEL Identified Technology Development Opportunities Prioritization. Washington, D.C.: Department of Energy. (DOE, undated b).

Department of Energy. 1992. Hanford Strategic Plan. Richland, WA: Department of Energy. (DOE, 1992).

Department of Energy. 1993. A New Approach to Environmental Research and Technology Development at DOE: Report of the Working Group (Viewgraphs). Washington, D.C.: Department of Energy. December 14. (DOE, 1993a).

Department of Energy 1993. EM Progress. Summer 1993. Washington, D.C.: Department of Energy. (DOE, 1993b).

Department of Energy. 1994. A New Approach to Environmental Research and Technology Development at the U.S. Department of Energy: Action Plan. Washington, D.C.: Department of Energy. January 25. (DOE, 1994a).

Department of Energy. 1994. FY 1995 Technology Development Needs Summary. Washington, D.C.: Department of Energy, Office of Environmental Management. March. (DOE, 1994b).

Department of Energy. 1995. Draft Technical Task Plan Proposal Evaluation Criteria--DG-24-95. Idaho Falls, ID: Department of Energy, Idaho National Engineering Laboratory. (DOE, 1995a).

Department of Energy. 1995. EM SSAB Recommendation on the EM FY-98 Integrated Priority List and RDS Rankings. Washington, D.C.: Department of Energy. November 15. (DOE, 1995b).

Department of Energy. 1995. Estimating the Cold War Mortgage: The 1995 Baseline Environmental Management Report, Vols. 1-3. Washington, D.C.: Department of Energy, Office of Environmental Management. March. (DOE, 1995c).

Department of Energy. 1995. INEL EM Prioritization IPT Report. Idaho Falls, ID: Department of Energy, Idaho National Engineering Laboratory. December. (DOE, 1995d).

Department of Energy. 1995. OST Background Series: 2: Final Report of the Task Force on Alternative Futures for the Department of Energy National Laboratories (Galvin Report). Washington, D.C.: Department of Energy. February 1. (DOE 1995e).

Department of Energy. 1995. Scoring Instructions for INEL EM Prioritization. Idaho Falls, ID: Department of Energy, Idaho National Engineering Laboratory. December. (DOE, 1995f).

Department of Energy. 1995. Site Needs Assessment Report. Washington, D.C.: Department of Energy. (DOE, 1995g).

Department of Energy. 1996. Attachment C--EM SSAB-INEL Meeting Minutes: Recommendation: Fiscal Year 1998 Prioritization of Environmental Management Activities. Washington, D.C.: Department of Energy. January 17. (DOE, 1996a).

Department of Energy. 1996. Charter of the Site Technology Coordination Group (STCG) for the Idaho National Engineering Laboratory. Idaho Falls, ID: Department of Energy. (DOE, 1996b).

Department of Energy. 1996. Decontamination and Decommissioning Focus Area Annual Report 1996. Washington, D.C.: Department of Energy. (DOE, 1996c).

Department of Energy. 1996. Efficient Separations and Processing Crosscutting Program Review Process. Washington, D.C.: Department of Energy. (DOE, 1996d).

Department of Energy. 1996. Environmental Management Fiscal Year 1998 Integrated Budget Prioritization. Washington, D.C.: Department of Energy. (DOE, 1996e).

Department of Energy. 1996. Evaluation Plan: Large-Scale Demonstration Projects, Decontamination and Decommissioning Focus Area. Washington, D.C.: Department of Energy. May 30. (DOE, 1996f).

Department of Energy. 1996. Final Meeting Minutes, January 16-17, Meeting of the Environmental Management Site Specific Advisory Board--Idaho National Engineering Laboratory. Held at the Centre on the Grove in Boise, Idaho: Department of Energy. (DOE, 1996g).

Department of Energy. 1996. INEL Identified Technology Development Needs Prioritization. Washington, D.C.: Department of Energy. (DOE, 1996h).

Department of Energy. 1996. INEL Identified Technology Development Opportunities Prioritization. Washington, D.C.: Department of Energy. (DOE, 1996i).

Department of Energy. 1996. Leveling the Playing Field. Washington, D.C.: Department of Energy. July 3. (DOE, 1996j).

Department of Energy. 1996. LITCO Director and DOE-ID Program Manager Approval of INEL EM Program, FY97 Murder Board Priority List, ADS Baseline Rev. 4. Idaho Falls, ID: Department of Energy. (DOE, 1996k).

Department of Energy. 1996. Mixed Waste Focus Area Technology Development Requirements Document, Chemical Oxidation. Washington, D.C.: Department of Energy. July 30. (DOE, 1996l).

Department of Energy. 1996. Mixed Waste Focus Area Technology Development Requirements Document, Mercury Stabilization. Washington, D.C.: Department of Energy. August 19. (DOE, 1996m).

Department of Energy. 1996. Mixed Waste Focus Area Technology Development Requirements Document, Mercury Amalgamation. Washington, D.C.: Department of Energy. July 30. (DOE, 1996n).

Department of Energy. 1996. Office of Science and Technology Decision Process Standard Operating Procedures (Draft). Washington, D.C.: Department of Energy. December 11. (DOE, 1996o).

Department of Energy. 1996. PI Requirements and Deliverables for the OST Technology Decision Process Procedures. Washington, D.C.: Department of Energy. (DOE, 1996p).

Department of Energy. 1996. Pipe Explorer™ System: Innovative Technology Summary Report. Hazardous Waste Remedial Actions Program. Oak Ridge, TN: Hazardous Waste Remedial Actions Program. April. (DOE, 1996q).

Department of Energy. 1996. Priority-Setting/Ranking Criteria. Washington, D.C.: Department of Energy. December 9-10. (DOE, 1996r).

Department of Energy. 1996. Robotics Crosscutting Program: Technology Summary. Washington, D.C.: Department of Energy, Office of Science and Technology. August. (DOE, 1996s).

Department of Energy. 1996. Robotics Technology Development Program Major Thrusts/Major Milestones Micro TTPs Fiscal Year 1997. Albuquerque, NM: Department of Energy, Albuquerque Operations Office. (DOE, 1996t).

Department of Energy. 1996. Tanks Focus Area Multiyear Program Plan FY97-FY99. Richland, WA: Department of Energy, Pacific Northwest Laboratory. August. (DOE, 1996u).

Department of Energy. 1996. Technology and Science Needs Summary (Draft). Washington, D.C.: Department of Energy. (DOE, 1996v).

Department of Energy. 1996. Technology Decision Process Pilot Review at Gate 4 of Subsurface Contaminant Focus Area Projects, ACT*DE*CON, Alternative Landfill Cover. Oak Ridge, TN: Department of Energy. October 28-30. (DOE, 1996w).

Department of Energy. 1996. Technology Needs Statements. Washington, D.C.: Department of Energy. (DOE, 1996x).

Department of Energy. 1996. Technology Needs/Opportunities Statement Outline (Hanford Site). Richland, WA: Department of Energy. December 13. (DOE, 1996y).

Department of Energy. 1996. The ESP Project Priority Procedure (Predecisional Draft). Washington, D.C.: Department of Energy. (DOE, 1996z)

Department of Energy. 1996. Decontamination and Decommissioning Focus Area Results of the National Needs Assessment, July 24-25. 1996. Washington, D.C.: Department of Energy. (DOE, 1996aa).

Department of Energy. 1997. 1997 D&D Needs. Washington, D.C.: Department of Energy. (DOE, 1997a).

Department of Energy. 1997. Accelerating Cleanup: Focus on 2006 (Discussional Draft). Washington, D.C.: Department of Energy, Office of Environmental Management. June. (DOE, 1997b).

Department of Energy. 1997. D&D Focus Area Quarterly Report, January-March 1997 Activity. Morgantown, WV: Department of Energy, Federal Energy Technology Center. (DOE, 1997c).

Department of Energy. 1997. Efficient Separations and Processing Crosscutting Program's (ESP-CP) presentation to the National Academy of Sciences, Committee for Environmental Management Technologies, Decision Making Subcommittee on June 6. (DOE, 1997d).

Department of Energy. 1997. Implementation Proposal for Center for Acquisition and Business Excellence. Morgantown, WV: Department of Energy, Federal Energy Technology Center. February 13. (DOE, 1997e).

Department of Energy. 1997. Information from the July 1997 PEG Meeting in Gaithersburg. Washington, D.C.: Department of Energy. (DOE, 1997f).

Department of Energy. 1997. New DOE-OST Deployment Initiative. Washington, D.C.: Department of Energy. (DOE, 1997g).

Department of Energy. 1997. OST Background Series: 3: COE Reports Relevant to the Office of Science and Technology. Washington, D.C.: Department of Energy. April. (DOE, 1997h).

Department of Energy. 1997. OST Process for Prioritization (Predecisional Draft). Washington, D.C.: Department of Energy. July 14. (DOE, 1997i).

Department of Energy. 1997. Strategic Plan. Washington, D.C.: Department of Energy. September. (DOE, 1997j).

Department of Energy. 1997. Technology Management Review. Washington, D.C.: Department of Energy, Office of Science and Technology. July 8. (DOE, 1997k).

Department of Energy. 1997. Technology Needs Statements. Washington, D.C.: Department of Energy. (DOE, 1997l).

Department of Energy. 1998. Accelerating Cleanup: Paths to Closure. Washington, D.C.: Department of Energy, Office of Environmental Management. June. (DOE, 1998a).

Department of Energy. 1998. Accelerating Cleanup: Paths to Closure (Draft). Washington, D.C.: Department of Energy. February. (DOE, 1998b).

Department of Energy. undated. Management Plan. Richland, WA: Department of Energy. (DOE, undated-a).

Department of Energy. undated. Mixed Waste Subgroup Technology Needs Scoring Sheet. Richland, WA: Department of Energy. (DOE, undated-b).

Department of Energy. undated. Problem Prioritization Criteria (Sample). Richland, WA: Department of Energy. (DOE, undated-c).

Department of Energy. undated. Tank Technology Needs Scoring Sheet (Sample). Richland, WA: Department of Energy. (DOE, undated-d).

Department of Energy, Environmental Protection Agency, Washington State Department of Ecology, Pacific Northwest National Laboratory, Tri-Cities Commercialization Partnership. 1996. Hanford Technology Deployment Center Program Plan. Richland, WA: Department of Energy. August. (DOE, et al. 1996).

E.I. Du Pont De Nemours and Company. 1993. Corporate Environmentalism: 1993 Progress Report. DuPont Printing and Publishing. (DuPont, 1993).

Ecology and Environment, Inc. 1996. Overview of Methods Used to Develop the Idaho National Engineering Laboratory: Identified Needs and Opportunities Prioritization Tables (Draft). Idaho Falls, ID: Ecology and Environment, Inc. November 4. (Ecology and Environment, 1996).

Edwards, Allen Louis. 1954. Statistical Methods for the Behavioral Sciences. New York: Rinehart. (Edwards, 1954).

Epstein, Marc J. 1995. Measuring Corporate Environmental Performance. (Epstein, 1995).

Fishburn, P.C. 1964. Decision and Value Theory. New York: John Wiley & Sons. (Fishburn, 1964).

Frank, Clyde. 1997. Office of Science and Technology Update and Program Data Analysis. Presentation to Committee on June 25. (Frank, 1997a).

Frank, Clyde. 1997. Presentation to the Committee on March 20, 1997. Washington, D.C.: Department of Energy. (Frank, 1997b).

Frey, Jeff, and Tom Brouns. 1997. Tanks Focus Area Process for Program Development Needs Identification & Technical Response Development. Richland, WA: Department of Energy, Tanks Focus Area. January 9. (Frey and Brouns, 1997).

Frolio, Nancy. 1996. Technology Decision Process Pilot Program. Washington, D.C.: Department of Energy. Memorandum to James Wright and Carl Bauer. November 14. (Frolio, 1996).

Gas Research Institute. 1996. Gas Research Institute: 1997-2001 Research and Development Plan and 1997 Research and Development Program. Chicago, IL: GRI. April. (GRI, 1996a).

Gas Research Institute. 1996. Results of Appraisal of GRI: 1997-2001 R&D Plan. Chicago, IL: GRI. April. (GRI, 1996b).

Gas Research Institute. 1997. Basic Research (1996 Results). Chicago, IL: GRI. June. (GRI, 1997a).

Gas Research Institute. 1997. Gas Research Institute: 1998-2002 Research and Development Plan and 1998 Research and Development Program. Chicago, IL: GRI. April. (GRI, 1997b).

Gas Research Institute. 1997. Results of Appraisal of GRI: 1998-2002 R&D Plan. Chicago, IL: GRI. June. (GRI, 1997c).

Gas Research Institute. 1998. Gas Research Institute 1999 to 2003 Plan and 1999 Research and Development Program. Chicago, IL: GRI. (GRI, 1998).

General Accounting Office. 1992. Cleanup Technologies. Washington, D.C.: General Accounting Office. (GAO, 1992).

General Accounting Office. 1994. Management Changes Needed to Expand Use of Innovative Cleanup Technologies: Report to the Secretary of Energy. Washington, D.C.: General Accounting Office. August. (GAO, 1994).

General Accounting Office. 1996. Energy Management: Technology Development Program Taking Action to Address Problems. Washington, D.C.: General Accounting Office. July. (GAO, 1996).

General Accounting Office. 1998. Nuclear Waste: Further Actions Needed to Increase Use of Innovative Cleanup Technologies. Report to Congressional Committees. Washington, D.C.: General Accounting Office. September. (GAO, 1998).

Hammond, John S., Ralph L. Keeney, and Howard Raiffa. 1999. Smart Choices: A Practical Guide to Making Better Decisions. Boston, MA: Harvard Business School Press. (Hammond, Keeney, and Raiffa, 1999).

Harness, Jerry. 1997. Presentation to the Committee. Washington, D.C.: Department of Energy. (Harness, 1997).

Hart, Paul, 1997. Decision Making within the Decontamination and Decommissioning Focus Area. Presentation to the Committee May 8-9. (Hart, 1997).

Heeb, Michael. 1996. Implementation of New Peer Review Policy. Washington, D.C.: Department of Energy. Memorandum for Distribution. November 18. (Heeb, 1996).

Holmes, J.D. 1994. White Water Ahead: Eastman Prepares for Turbulent Times. Industrial Research Institute: Research Technology Management. (Holmes, 1994).

Holt, Mark and Jeff Day. 1997. Memorandum, Congressional Research Service: Mark Holt & Jeff Day to Senate Appropriations Committee regarding Environmental Technology Development at DOE. Washington, D.C.: Congressional Research Service. (Holt and Day, 1997).

Howard, Ronald A. 1966. Dynamic Programming and Markov Processes. Cambridge, MA: Technology Press of MIT. (Howard, 1966).

Howard, R.A., and J.E. Matheson (eds.). 1983. The Principles and Applications of Decision Analysis. Palo Alto, CA.: Strategic Decisions Group. (Howard and Matheson, 1983).

Hyde, Jerry. 1996. Correspondence from Jerry Hyde to Carl Bauer and Tom Anderson regarding D&D National Needs. August 1. (Hyde, 1996).

Keeney, Ralph L. Value-Focused Thinking: A Path to Creative Decision-Making. Cambridge, MA: Harvard University Press. (Keeney, 1992).

Lander, Lynn, David Matheson, Michael M. Menke, and Derek L. Ransley. 1995. Improving the R&D Decision Process. Industrial Research Institute: Research-Technology Management (January-Februrary). (Lander, Matheson, Menke and Ransley, 1995).

Lankford, Mac. 1997. Technology Integrations Systems Applications Programs--Domestic Programs--International Programs. Presentation to the Committee on November 10, Washington, D.C. Department of Energy. (Lankford, 1997).

Levary, Reuven, and Neil E. Seitz. 1990. Quantitative Methods for Capital Budgeting. Cincinnati, OH: South-Western Publishing Company. (Levary and Seitz, 1990).

LEAF v. Hodell. 586 Federal Supp. 1163.

Love, Betty R. 1997. Correspondence to Steven J. Bossart regarding Consensus Reports for the Review Panel of RSI. Institute for Regulatory Science. April, 9. (Love, 1997).

Luce, R.D., and H. Raiffa. 1958. Games and Decisions. New York: John Wiley & Sons. (Luce and Raiffa, 1958).

March, James G. 1994. A Primer on Decision Making: How Decisions Happen. New York, NY: The Free Press. (March, 1994).

March, James G. 1999. The Pursuit of Organizational Intelligence. Malden, MA: Blackwell Publishers, Inc. (March, 1999).

Markel, Ken. 1997. Partnering with the Private Sector: The Key to Future Federal R&D Investment. Presentation to Committee on May 8-9. (Markel, 1997).

Martin, H. Dale. 1994. Environmental Planning: Balancing Environmental Commitments with Economic Realities. Presentation at the Proceedings of the Environmental Management in a Global Economy (GEMI'94) Conference. (Martin, 1994).

Martin, H. Dale. 1997. Environmental Planning--Balancing Environmental Commitments with Economic Realities. Speech before the Southeastern Electrical Exchange, Charlotte, NC: June 26. (Martin, 1997).

Martino, J. 1995. R&D Project Selection. New York: John Wiley & Sons. (Martino, 1995).

Matalucci, Rudolph V., Charlene Esparaza-Baca, and Richard D. Jimenez. 1995. Characterization, Monitoring, and Sensor Technology Catalogue. Albuquerque, NM: Sandia National Laboratories. December. (Matalucci, Esparanz-Baca, and Jimenez, 1995).

Matheson, D., and J. Matheson. 1998. The Smart Organization. Cambridge, MA: Harvard Business School Press. (Matheson and Matheson, 1998).

Matheson, James E. and Michael M. Menke. 1994. Using Decision Quality Principles to Balance Your R&D Portfolio. Industrial Research Institute: Research-Technology Management (November-December). (Matheson and Menke, 1994)

Matheson, David, James E. Matheson, and Michael M. Menke. 1994. Making Excellent R&D Decisions. Industrial Research Institute: Research-Technology Management. (Matheson, Matheson, and Menke, 1994).

Mathur, John. 1997. Discussion with the Committee. Washington, D.C.: Department of Energy. June. (Mathur, 1997).

Menke, Michael M. 1994. R&D Quality Tools. Industrial Research Institute: Research-Technology Management. (Menke, 1994).

Menke, Michael M. 1997. Essentials of R&D Strategic Excellence. Industrial Research Institute: Research Technology Management. (Menke, 1997a).

Menke, Michael M. 1997. Managing R&D for Competitive Advantage. Research-Technology Management. (Menke, 1997b).

Myers, Joy. 1996. Correspondence to John Wilcynski regarding recommendations from the Environmental Management Site Specific Advisory Board. March 25. (Myers, 1996)..

National Environmental Technology Applications Center. 1995. Barriers to Environmental Technology Commercialization. Pittsburgh, PA: University of Pittsburgh Applied Research Center. April. (NETAC, 1995).

National Research Council. 1989. Nuclear Weapons Complex: Management for Health, Safety, and the Environment. Washington, D.C.: National Academy Press. (NRC, 1989).

National Research Council. 1994. Building Consensus Through Risk Assessment and Management of the Department of Energy's Environmental Remediation Program. Washington, D.C.: National Academy Press. (NRC, 1994).

National Research Council. 1995. Improving the Environment. Washington, D.C.: National Academy Press. (NRC, 1995a).

National Research Council. 1995. The Committee on Environmental Management Technologies Report for the Period Ending December 31, 1994. Washington, D.C.: National Academy Press. (NRC, 1995b).

National Research Council. 1996. Barriers to Science: Technical Management in the Department of Energy Environmental Remediation Program. Washington, D.C.: National Academy Press. (NRC, 1996a).

National Research Council. 1996. Environmental Management Technology-Development Program at the Department of Energy 1995 Review. Washington, D.C.: National Academy Press. (NRC, 199b).

National Research Council. 1996. The Hanford Tanks: Environmental Impacts and Policy Choices. Washington, D.C.: National Academy Press. (NRC, 1996c).

National Research Council. 1996. Understanding Risk: Informing Decisions in a Democratic Society. Washington, D.C.: National Academy Press. (NRC, 1996d).

National Research Council. 1997. Building an Effective Environmental Management Science Program: Final Assessment. Washington, D.C.: National Academy Press. (NRC, 1997a).

National Research Council. 1997. Peer Review in the Department of Energy-Office of Science and Technology: Interim Report. Washington, D.C.: National Academy Press. (NRC, 1997b).

National Research Council. 1998. A Review of Decontamination and Decommissioning Technology Development Programs at the Department of Energy. Washington, D.C.: National Academy Press. (NRC, 1998a).

National Research Council. 1998. Peer Review in Environmental Technology Development Programs: The Department of Energy--Office of Science and Technology. Washington, D.C.: National Academy Press. (NRC, 1998b).

National Research Council. 1999. An End State Methodology for Identifying Technology Needs for Environmental Management: With an Example from the Hanford Site Tanks. Washington, D.C.: National Academy Press. (NRC, 1999a).

National Research Council. 1999. Evaluating Federal Research Programs: Research and the Government Performance and Results Act. Washington, D.C.: National Academy Press. (NRC, 1999b)

Paladino, Joseph, and Brian Fox. undated. A Framework for R&D Planning in EM (Draft). Prepared for the TD Council and EM Focus Group. (Paladino and Fox, undated).

Paladino, Joseph, and Paul Longworth. 1995. Maximizing R&D Investments in the Department of Energy's Environmental Cleanup Program. Technology Transfer (December). pp. 96-107. (Paladino and Longworth, 1995).

Pratt, J., H. Raiffa, and R. Schlaifer. 1965. Introduction to Statistical Decision Theory. New York: McGraw-Hill. (Pratt, Raiffa, and Schaifler, 1965).

President's Committee of Advisors on Science and Technology. 1997. Report to the President on Federal Energy Research and Development for the Challenges of the Twenty-First Century. Washington, D.C.: President's Committee of Advisors on Science and Technology, Panel on Energy Research and Development. November. (PCAST, 1997).

Purdy, Caroline. 1997. Decision Making Review: Characterization, Monitoring, and Sensor Technologies. Presentation to Committee. (Purdy, 1997).

Raiffa, H. 1968. Decision Analysis. Redding, Mass.: Addison-Wesley. (Raiffa, 1968).

Rezendes, Victor S. 1997. Cleanup Technology: DOE'S Program to Develop New Technologies for Environmental Cleanup. Testimony before the United States House of Representatives Subcommittee on Oversight and Investigations. Washington, D.C.: General Accounting Office. (Rezendes, 1997).

Russell, Milton, E. William Colglazier, and Mary English. 1991. Hazardous Waste Remediation: The Task Ahead. Knoxville, TN: Waste Management Research and Education Institute, University of Tennessee. December. (Russell, Colglazier, and English).

Russo, J. Edward, and Paul Schoemaker. 1989. Decision Traps: Ten Barriers to Excellent Decision Making and How to Overcome Them. New York, NY: Simon & Schuster. (Russo, and Shoemaker, 1989).

Saaty, Thomas L. 1994. Fundamental of Decision Making and Priority Theory: With the Analytic Hierarchy Process, Vol. VI. Pittsburgh, PA: RWS Publications. (Saaty, 1994).

Saaty, Thomas L. 1996. Decision Making with Dependence Feedback: The Analytic Network Process. Pittsburgh, PA: RWS Publications. (Saaty, 1996).

Saaty, Thomas L., and Luis G. Vargas. 1998. Diagnosis with Dependent Symptoms: Bayes Theorem and The Analytic Hierarchy Process. Operations Research, Vol. 46 (4): July-August. INFORMS. pp. 491-502. (Saaty and Vargas, 1998).

Saaty, Thomas L., and Luis G. Vargas. 1993. Experiments on Rank Preservation and Reversal in Relative Measurement. Mathematical and Computer Modeling, Vol. 17 (4/5): Pergamon Press, Ltd. pp. 13-18. (Saaty and Vargas, 1993).

Saget, Shannon. 1996. Summary--Hanford STCG Prioritization Methodology. Richland, WA: Department of Energy. (Saget, 1996).

Scott, James C. 1998. Seeing Like A State: How Certain Schemes to Improve the Human Condition Have Failed. New Haven, CT: Yale University Press. (Scott, 1998).

Simon, Herbert A. 1997. Administrative Behavior: A Study of Decision-Making Processes in Administrative Organizations (fourth edition). New York, NY: The Free Press. (Simon, 1997).

Simon, Herbert A., Donald W. Smithburg, and Victor A. Thompson. 1991. Public Administration. New Brunswick, NJ: Transaction Publishers. (Simon, Smithburg, and Thompson 1991).

Snelling, Robert N., and Susan Wood. 1996. Correspondence to Clyde Frank regarding Transmittal of Vitrification Systems Information Package. May 1. (Snelling and Wood, 1996).

Stigum, Bernt P., and Fred Wenstop. 1983. Foundations of Utility and Risk Theory with Applications. Dordrecht, Holland: D. Reidel Publishing Company. (Stigum and Wenstop, 1983).

Surles, Terry. 1997. Letter to Alvin Alm with viewgraphs entitled "Review Recommendations for Office of Science and Technology Policy". May 15. (Surles, 1997).

Tribus, Myron. 1969. Rational Descriptions, Decisions, and Designs. New York: Pergamon Press. (Tribus, 1969).

Tuchman, Barbara. 1992. The March of Folly: From Troy to Vietnam. New York, NY: Ballantine Books. (Tuchman, 1992).

Tversky, Amos, and Itamar Simonson. 1993. Context-Dependent Preferences. Management Science, Vol. 39 (10): October. Institute of Management Sciences. pp. 1179-1189. (Tversky and Simonson, 1993).

Tversky, Amos, Paul Slovic, and Daniel Kahneman. 1990. The Causes of Preference Reversal. The American Economic Review, Vol. 80 (1): March. pp. 204-217. (Tversky, Slovic, and Kahneman, 1990).

United States. Army Corps of Engineers. 1997. Project EM Task Force Phase 1 Report. Washington, D.C.: U.S. Army Corps of Engineers. February. (U.S. Army Corps of Engineers, 1997).

United States Congress. 1989. National Defense Authorization Act for Fiscal Years 1990 and 1991. P.L. 101-189, Sec. 3141 (H.R. 2461). Washington, D.C.: U.S. Government Printing Office. (US Congress, 1989).

United States Congress, Senate. 1992. Energy and Water Development Appropriations Bill, 1993. Report to accompany H.R. 5373. 102d Congress, 2d session, pp. 138-139. July 27. (U.S. Congress, 1992).

United States Congress, House of Representatives. 1994. Energy and Water Development Appropriations Bill, 1995. Report to accompany H.R. 4506. 103d Congress, 2d session, pp.76-77, 100. May 26. (U.S. Congress, 1994).

United States Congress, House of Representatives. 1995. Energy and Water Development Appropriations Bill, 1996. Report to accompany H.R. 1905. 104th Congress, 1st session, pp. 77-78. June 20. (U.S. Congress, 1995a).

United States Congress, Senate. 1995. Energy and Water Development Appropriations Bill, 1996. Report to accompany H.R. 1905. 104th Congress, 1st session, pp. 110-111. July 27. (U.S. Congress, 1995b).

United States Congress, House of Representatives, Committee on Commerce, Subcommittee on Oversight and Investigations. 1997. Summary of Congressional Testimony on Department of Energy/Office of Science and Technology. Washington, D.C.: Committee on Commerce. May 7. (U.S. Congress, 1997).

United States Congress, Office of Technology Assessment. 1986. Research Funding as an Investment: Can We Measure the Returns?--A Technical Memorandum. Washington, D.C.: Government Printing Office. April. (U.S. Congress, 1986).

Von Neumann, J., and O. Morgenstern. 1953. Theory of Games and Economic Behavior. Revised edition. Princeton, N.J.: Princeton Univ. Press. (Von Neumann and Morgenstern, 1953).

Walker, Jef. 1997. Current Status and Future Directions of Science and Technology in the EM Program. Presentation to the National Academy of Sciences Committee on Peer Review. (Walker, 1997a).

Walker, Jef. 1997. New OST Process for Prioritization: Making OST Decisions More Responsive to the DOE Sites through the 2006 Plan. Paper presented to Committee on November 10. (Walker, 1997b).

Wang, Paul. 1997. Correspondence from Paul Wang to CEMT Committee Members. Santa Barbara, CA: Special Technologies Laboratory. May 23. (Wang, 1997).

Wengle, John. 1997. E-mail Correspondence from John Wengle to Carolyn Davis regarding NAS request for information. October 30. (Wengle, 1997).

Williams, R. Eric. 1996. Correspondence to Julie Conner regarding Mixed Waste Focus Area High Temperature Melter Strategy Recommendations for FY97. May 31. (Williams, 1996).

Wolf, Stanley. 1997. E-mail Correspondence from Stanley Wolf to Carolyn Davis regarding reply to ORO Information Request from NAS. October 24. (Wolf, 1997).

Wright, Jim. 1996. Priority-Setting/Ranking Criteria Presentation to the Committee. National Academy of Sciences, Committee on Environmental Management Technologies, Decision-Making Process for Selection and Prioritization Panel. (Wright, 1996).

Yarbrough, Linton. 1997. Robotics Technology Development Program. Presentation to Committee on June 6. (Yarbrough, 1997).

Zeleny, Milan. 1982. Multiple Criteria Decision Making. New York, NY: McGraw-Hill Book Company. (Zeleny, 1982).

Appendix A

Summary of Past NRC Reports on Decision Making
in DOE-EM and OST

Several previous NRC reports have noted issues in the DOE-EM program that are relevant to OST decision making. The principal findings, conclusions, and recommendations of nine such reports are summarized below.

NRC REPORTS PERTINENT TO DOE-EM BUT NOT ADDRESSED TO OST IN PARTICULAR

1. The report *Nuclear Weapons Complex: Management for Health, Safety, and the Environment* (NRC, 1989), written at the beginning of the DOE-EM program, made several observations, recommendations, and conclusions reproduced here because of their relevance to decision making.

On the Decision-Making Process

Observations:

- Decisions need to be pushed downward in the hierarchy of DOE,
- There is a tendency to push things upward, especially during budget deliberations,
- Lines of authority should be clear, simple, and unambiguous,
- Headquarters must provide clear guidance to the field offices and contractors if objectives are to be achieved.

"Conclusion: Many decisions are now unnecessarily deferred by staff to higher management levels, sometimes creating delay and paralysis in decision making" (NRC, 1989, p. 22).

"Recommendation: The Department should strengthen its management structure by delegating authority and responsibility for the initial resolution of issues to the lowest possible management levels, subject to clear guidance and support from upper management" (NRC, 1989, p. 22).

On Exchange of Information Within the Complex

"Conclusion: Communications among organizations that confront common problems and efforts to focus the resources of the complex on finding solutions are inadequate" (NRC, 1989, p. 27).

"Recommendation: The Department should work harder to overcome the natural impediments to the flow of information among contractors and to facilitate communication among the contractors and among DOE staff" (NRC, 1989, p. 29).

On Independent Technical Advice

"Conclusion: The Department is not aggressive enough in seeking the advice and counsel of experts from outside the weapons complex" (NRC, 1989, p. 29).

"Recommendation: The Department should aggressively seek outside advice, from the ACNFS [Advisory Committee on Nuclear Facility Safety] and other sources, with regard to the many technical issues that it confronts in the operation of the weapons complex" (NRC, 1989, p. 30).

On Setting Standards and Priorities Across the Complex

"Conclusion: There is a need to develop and apply a scientifically credible scheme to aid in making decisions about appropriate cleanup standards and priorities for performing remediation in the face of resource limitations" (NRC, 1989, p. 38).

"Recommendation: The Department should seek to achieve site-specific cleanup standards. Consistent risk assessment methodologies should be used to bring scientific information into decisions regarding extent of cleanup, cleanup methodologies, and priorities for environmental restoration" (NRC, 1989, p. 41).

"Recommendation: To the greatest extent practicable, DOE should incorporate risk assessment as a guiding principle in developing an NPS [National Priority System]" (NRC, 1989, p. 42).

On Characterization of Contaminated Sites

"Conclusion: Intensified sampling to describe the extent and nature of contamination, as well as hydrogeology and ecology, is necessary to guide cleanup, isolation, or restoration activities in a timely manner. Improved data management will assist in the retrieval and analysis of the massive amount of information collected" (NRC, 1989, p. 42).

"Recommendation: Each installation should develop a comprehensive data base of environmental information, one that will allow the data to be accessed and used for a variety of purposes related to remediation of contaminated sites within the installation. The structure and content of the database should be consistent across the complex. DOE should also insist that each installation develop a plan to acquire the data necessary to improve understanding of the installation-wide geology, hydrogeology, and land use" (NRC, 1989, p. 43).

   2. The report *Building Consensus Through Risk Assessment and Management of the Department of Energy's Environmental Remediation Program* (NRC, 1994) noted that

   • the lack of trust in DOE and site operators is an impediment to the consensus and decision process;

- many interested parties (e.g., stakeholders) need to be involved from the beginning (i.e., at planning stages);
- any such process to solicit their involvement should be open, clear, equitable, and inclusive; and
- the absence of complete information should not be an excuse for lack of progress in site remediation.

The report commended risk assessment as "a highly desirable component of the remediation decision-making process" (NRC, 1994, p. 3), and said it was "one of a number of elements in the decision-making process and should not be treated as the only one" (NRC, 1994, p. 4). A comprehensive risk assessment process is feasible, desirable, and essential in consensus building for key decisions (NRC, 1994, p. 4).

3. The report *Improving the Environment: An Evaluation of DOE's Environmental Management Program* (NRC, 1995a) noted that the DOE-EM program had the following needs:

- more specific goals;
- a process for prioritizing tasks that would include risk assessment and cost-benefit analysis;
- a peer-reviewed process for technology selection and development that is responsive to the needs of those implementing the remediation; and
- incentives for successful implementation.

These needs are contained in the report's principal recommendation:

This committee believes getting on with the task, whichever definition one uses, will be accomplished most effectively by implementing a process for decision-making and accountability that includes

- Having a more specific set of goals for the program
- A process for prioritizing tasks which includes among its tools risk assessment (which should consider the perspectives and values of stakeholders as recommended in *Building Consensus* (NRC, 1994) and cost-benefit analysis.
- A peer-reviewed remediation and waste-minimization technology selection and development process that is responsive to the needs of those implementing the remediation
- An overall organizational and management structure which both provides an opportunity for stakeholder input in each of the above activities and provides incentives for stakeholders and federal and contract workers to implement these activities of the Environmental Management Program successfully. (NRC, 1995a, pp. 10-11)

4. The report *Barriers to Science: Technical Management of the Department of Energy Environmental Remediation Program* (NRC, 1996a) noted the following seven problems:

1. planning that is driven by organizational structures rather than by problems to be solved;
2. commitments that are made without adequately considering technical feasibility, cost,   and/or schedule;
3. an inability to look at more than one alternative at a time;
4. priorities that are driven by narrow interpretations of regulations rather than the regulatory purpose of protecting public health and the environment;
5. the production of documents as an end in itself, rather than as a means to achieve a goal;

6. a lack of organization coordination; and
7. a "not-invented-here" syndrome at sites.

The report noted the following attributes of the DOE-EM program that were contributing concepts:

- lack of progress toward stated goals,
- a general inefficiency,
- goals that were not correctly formatted, and
- bad incentive programs.

The report deduced the following problem with the DOE-EM program (NRC, 1996a, p. 1):

> The committee observes a common pattern of behavior in these problems: What happens is driven too often by the internal needs of the organizational units charged with the remediation work rather than by the overall goal of environmental remediation. Efforts to remedy this situation must involve not only the Department of Energy, but also external stakeholders who have influenced its ways of doing business, including Congress, involved states, and the public.

This problem is an organizational one, of technical management (NRC, 1996a, p. 19). External forces, as well as DOE culture, are contributing factors. An "aversion to decision-making" (NRC, 1996a, p. 19) is mentioned.

5. The report *The Hanford Tanks: Environmental Impacts and Policy Choices* (NRC, 1996c) reviewed alternative remediation approaches to the Hanford tanks, and noted that (1) significant uncertainties exist that limit current knowledge and that therefore (2) a phased decision strategy to consider multiple alternatives is called for.

6. *Understanding Risk: Informing Decisions in a Democratic Society* (NRC, 1996d) emphasizes the role of risk characterization as a systematic analysis that includes participation by interested parties and as "a *decision-driven activity*, directed toward informing choices and solving problems" (NRC, 1996d, p. 2). The report noted the importance of early and total stakeholder-public involvement and the issue of public trust (or mistrust) in dealing with a government program.

7. The report *Evaluating Federal Research Programs: Research and the Government Performance and Results Act* (NRC, 1999b) examined research programs in federal institutions, including the Department of Energy, and offered recommendations as to how to assess the value of such programs. One recommendation was that expert review (an enhanced form of peer review that calls on individuals with sufficient expertise in the field to be considered to review the program) is the most effective method of reviewing federally funded research programs, concentrating on the following three indicators:

I. *quality of research programs*
II. *relevance to agency mission and*
III. *international benchmarking.*

NRC REPORTS PREPARED FOR OST

8. *The Committee on Environmental Management Technologies Report for the Period Ending December 31, 1994* (NRC, 1995b) noted, among others, the following characteristics of and advice to the OST program (as of December 1994):

- a lack of quantifiable end goals, cleanup levels, and criteria (to define technology development specifications and cleanup end points);
- environmental remediation should be an experiment, with flexibility, integration of efforts, and feedback;
- long-term work, particularly basic research, should continue, because the cleanup program will take more than 30 years;
- it is recommended that R&D focus on identified problems at particular sites;
- the final disposal destination of wastes should be considered in remediation decisions;
- the concept of waste minimization is praised as a criterion for technology development projects;
- the development of technology backups to the current baseline methods is lauded;
- vagueness is a noted characteristic of both

    1. numerical data and
    2. the bases and assumptions underlying the derivation of calculated numbers;

- the committee questions DOE's abilities to

    1. explore alternative technologies, after being locked into legal regulatory agreements for long-term (multi-decade) cleanups, and
    2. work with regulators in bringing new information to bear in renegotiating these agreements;

- reasonable guidelines should be worked out between DOE and EPA on ground water cleanups for which it is unrealistic to expect to achieve drinking water standards;
- the committee urged organizational continuity and stated that Focus Areas deserve a chance to see if they work well; and
- redundant or duplicative technology development projects should be eliminated.

The report noted three priority issues for OST:

1. develop technologies that minimize secondary waste,
2. develop backup technologies, and
3. correct the vague criteria and apparent lack of justification that characterize the selection of site remediation measures, the desirable goals of remediation efforts, and the use of such information in technology development decisions.

The report concludes (NRC, 1995b, p. 9):

A culture change in technology development is needed at DOE-EM, where established programmatic and funding patterns are attuned to meeting the exigencies of site emergencies and external pressures. The committee hopes that the emphasis will turn to more systematic evaluation, including the use of risk assessment, to allow a more rational approach to technology development and site remediation.

9. The report *Environmental Management Technology-Development Program at the Department of Energy 1995 Review* (NRC, 1996b), quoted below, gave recommendations that relate to the processes of goal setting, program planning, peer review, information gathering, and other activities that affect OST decision making. As expanded upon in the text, the Executive Summary noted (NRC, 1996b, p. 2):

Specific actions that DOE needs to take are

- develop and implement quantitative criteria by which technology-development efforts can be prioritized and success can be measured;
- carefully consider the waste streams (including those from remediation efforts to their eventual disposition) in determining adequate technology-development needs;
- systematically assess and document previous and current efforts to develop and apply technologies using the quantitative criteria mentioned above;
- apply effective, external peer review in the selection, evaluation, and prioritization of its projects; and
- improve its system for information gathering and documentation on technologies that are available and under development by other relevant organizations in the United States and abroad.

10. *Building an Effective Environmental Management Science Program: Final Assessment* (NRC, 1997a) provided a recognition of the general value of scientific research in areas relevant to DOE-EM cleanup problems. The report recommends that the newly formed Environmental Management Science Program develop a "science plan." Other recommendations pertain to processes for proposal evaluation and program management strategies to enhance the long-term effectiveness and credibility of the EMSP.

11. *Peer Review in the Department of Energy-Office of Science and Technology: Interim Report* (NRC, 1997b) provided recommendations on the implementation of OST's peer review system to review projects that have been funded by the program and have reached sufficient maturity to show results of the technology development efforts. The report stressed that peer reviews should be conducted by technical experts external to (and independent of) the OST program office and highlighted the issues of 1. defining the appropriate set of technical criteria to use in peer reviews and 2. defining the objectives of each peer review.

Appendix B

Site Technology Coordination Groups at Three Major DOE-EM Sites

Site Technology Coordination Groups are responsible for developing and prioritizing a list of site problems and technology needs based on the environmental management issues relevant to a specific site. A technology needs statement describes the nature, magnitude, environmental conditions, and technical performance requirements of an environmental restoration or remediation problem. The STCG is intended to be the primary interface between the site remediation programs and the national Focus Areas. The STCGs forward the site's technology needs list to the Focus Areas, where the technology needs are compiled and prioritized to address problems and needs at multiple sites in the context of national technology needs. This information is used to analyze and update the Focus Area portfolio of technologies and determine investment targets. If the Focus Area is aware of an available technology solution, this information is transmitted to the applicable STCG, where the site can implement the technology.

A significant additional role of STCGs is to coordinate all technology development activities at the site and facilitate implementation of new technologies site-wide to improve the performance of environmental remediation and waste management programs. This requires that the STCGs develop performance requirements for technology implementation, participate in site technology demonstrations, and evaluate the ability of implemented technologies to meet site needs. A corollary role of the STCGs is to review the Focus Area technology development activities to evaluate whether the Focus Areas adequately understand the sites' problems and respond to their respective needs and priorities. The formalized process used by the STCGs in identifying and prioritizing technology needs is intended to define technology needs and link technology users with the Focus Areas' efforts to promote the use of innovative technologies at multiple DOE sites.

The composition of STCGs differs somewhat from site to site. At most sites, STCG members include representatives of DOE, site contractors, and a variety of stakeholders, including government agencies, public interest groups, and Native American groups, whereas STCGs at some sites are comprised solely of DOE employees. Each STCG considers the regional setting, environmental problems, regulatory issues, legally binding commitments, and priorities of the individual site.

EM-30 and EM-40 work through the sites' Technical Program Officer (TPO) and STCG to identify technology needs and priorities. Technology need information includes site-specific descriptions of the nature of the problem, baseline technologies and costs, regulatory requirements and issues, environmental safety and health concerns, anticipated cleanup schedule, potential cost savings (mortgage reduction), end user performance requirements, and local cultural or stakeholder concerns. Information must also be provided regarding the applicability of the technology need to multiple DOE sites or to private industry. This information is important for the Focus Areas in their complex-wide prioritization of technology programs.

The STCGs compile information in a standardized format that provides a complete description of the technology requirements in terms of site conditions, operational requirements, and technical performance.

Information submitted to the Focus Areas is reviewed for completeness, and if needed, further information is requested through the TPO or STCG. Focus Areas evaluate the combined needs statements from all sites and combines them into Focus Area work packages, groupings of needs based on similarity in characteristics and program needs.

In principle, results of the work package evaluation and prioritization that form the basis for soliciting technology development proposals are then provided to the STCGs. The needs that have a high end user commitment and potential for deployment and implementation at multiple sites are more likely to be selected for funding. Some needs, however, are not justified because the baseline technology is determined to be adequate or more cost-effective, whereas other needs may have commercially available technology solutions. Other technology needs may be more effectively addressed in a different program or are not of high enough priority in *Accelerating Cleanup: Paths to Closure* (DOE, 1998a) to warrant further action.

SAVANNAH RIVER STCG

The STCG at the Savannah River Site (SRS) addresses the technology needs of the SRS line organizations (Environmental Restoration, High Level Waste, and Solid Waste). Since the SRS STCG is an oversight and coordinating group, the line organizations and DOE line management have responsibility for identifying technology needs and implementing new technologies. The STCG facilitates identification and prioritization of technology needs and attempts to obtain the resources to implement appropriate technologies. The STCG also provides for interaction with SRS stakeholders, including citizens, advisory boards, and government agencies, to maintain communication and obtain input regarding new technology development and implementation.

At SRS, the STCG is comprised primarily of Savannah River Operations Office (DOE-SR) personnel and is chaired by the TPO. Members of the STCG include Focus Area liaisons that serve as the primary interface between the line organizations at SRS and the national Focus Areas. Focus Areas are represented in the following six technology teams:

1. High-Level Waste in Tanks,
2. Environmental Restoration (Subsurface Contaminants Focus Area),
3. Waste Management/Site Treatment Plan (Mixed Waste Focus Area),
4. Decontamination and Decommissioning
5. Spent Fuel, and
6. Nuclear Focus Area Materials Stabilization (Plutonium Focus Area).

The STCG also includes representatives of DOE-SR groups that interface with regulatory agencies and other stakeholders, as well as representatives of the site contractor—the end user with the technology need. Supporting members include representatives of the Offices of Community Outreach, Economic Development, the Chief Counsel, and the Savannah River Technology Center.

The STCG Executive Committee, composed of DOE-SR senior managers, provides overall program guidance and direction, and is intended to serve as an interface with senior management to confirm site-wide priorities.

The STCG at SRS developed a needs identification format that has been adopted by other STCGs in the complex to document needs. Additionally, the Environmental Restoration Program has developed performance-based standards for evaluating the successful implementation of new technologies. Performance is gauged by the ability of the technology to reduce long-term costs (mortgage reduction) and to be deployed at multiple sites.

Each of the six Technology Teams develops a list of technology needs that are included in *Accelerating Cleanup: Paths to Closure* (DOE, 1998a) and are ranked according to different prioritization criteria and submitted to the STCG. In this process, a variety of both quantitative and nonquantitative criteria are used to prioritize needs. This allows individual technical teams to develop criteria relevant to their specific area of concern; however, it is difficult for the STCG to compare priorities among technical areas.

The needs lists developed by the SRS STCG are forwarded to the Citizen's Advisory Board (CAB) for review. The CAB, which has three subcommittees (Nuclear Materials Management, Environmental Restoration, and Future Use), is given the opportunity to make nonbinding recommendations. Stakeholders such as the CAB are able to review the prioritized list of technology needs; however, they do not have input in developing criteria or weights in the prioritization process.

Discussion with STCG representatives at SRS revealed an issue regarding lack of feedback in terms of the technology need requests submitted to the Focus Areas. In the Environmental Restoration (ER) Program, 15 technology need requests were submitted to EM-50 Subsurface Contamination Focus Area (SCFA, also known as the "Subcon" Focus Area) in 1996. When no response was received, the ER program submitted 12 proposals specifically identifying the technologies it requested to be funded by the SCFA. After no response was received again, this number was reduced to five proposals, none of which was funded. Apparently, the SCFA could not fund any new projects because of reduced budgets and was even required to reduce funding to existing, previously funded projects. Better communications between the STCG and the Focus Areas would significantly reduce such inefficient and time-wasting efforts in the needs identification and request process.

Other STCG-related issues included concerns regarding parochialism in the funding process and the need for greater involvement of the end user in the process. The latter is a common issue and one that reflects the problem in being able to deploy technologies at sites.

Idaho National Engineering and Environmental Laboratory (INEEL) STCG

The information below comes from documents supplied to the committee and from a site visit to Idaho Falls, Idaho, in November 1996.

Structure and Composition

The INEEL STCG is composed of two major units, the STCG Management Team and the STCG Working Group. The STCG Management Team consists of upper-level DOE and site contractor (Lockheed Martin) management, who approve the final STCG reports. The STCG Working Group is an integrating team, again composed of both DOE and contractor personnel that performs the work of generating documents for the Management Team to review.

Decision Process

The INEEL STCG work to prioritize technology needs was able to take advantage of other, immediately preceding efforts at similar tasks.

Resources Available to the STCG

One important resource was the LMITCO translation of all legally binding milestones and end points into a compilation of technology development requirements (Beller, 1997).

A second important resource was the Activity Data Sheet (ADS) baseline priority list, Revision. 4 (June 13, 1996). The ADSs define the scope of funded work on site, with technical deficiencies called out as decision units (DUs) (Ecology and Environment, 1996). The detailed description of each DU appears in a standardized Risk Data Sheet (RDS) format. Senior-level DOE and contractor (LMITCO) managers have ranked 169 DUs in importance to form the FY 1997 Murder Board Priority List to the ADS baseline.

A third resource was the result of work to develop the EM Integrated Site Priority List by combining the technology development requirements of Beller with the ADS baseline priority list and RDSs. These efforts are outlined below.

EM Integrated Site Priority List

An important input to the STCG process was the EM Integrated Site Priority List, developed by an integrated product team composed of DOE and contractor personnel. This team examined the DUs to consolidate and eliminate some,[1] and then to prioritize them using a multi-attribute decision analysis exercise.

Prioritization was done on two separate lists:

1. technology development *needs*, and
2. "alternative technology development *opportunities*" (Ecology and Environment, 1996, p. 2).

The statements in these lists sometimes combine several DUs into a broader statement of the technical need or opportunity.

In essence, the integrated product team's prioritization exercise on these two sets of statements assigned priorities to these statements, priorities that at the end differed from those originally assigned by the LMITCO contractor (DOE, undated a and b).

These decision units were assessed against many criteria, including

- vulnerabilities in cost and schedule: schedule and cost constraints were converted into a "can't be done" vulnerability, generated if the present schedule or funding profile is inadequate to meet a cleanup milestone; and
- determination of benefit: the benefit of doing a technical task in a cleanup operation was defined (using cost as an attribute) as the cost of doing the task minus the cost of not doing the task.

Development of Analytical Weighting Factors

Input from several sources, including stakeholders, was used to establish relative weights of different attributes. Questions designed to elicit comparisons were used to elucidate the relative weighting factors used in the multi-attribute decision analysis. Further details of this decision process methodology can be found in DOE (1995d, 1995f). The evolution of efforts to arrive at an integrated priority list was monitored by the active site specific advisory board (DOE, 1995b, 1996a, 1996e, 1996g, 1996a; Myers, 1996.).

---

[1]This involved combining the information in the list of DUs from the ADS with the technology development requirements from Beller (1997).

Output of the Process

The result of the multi-attribute utility analysis was a numerical score assigned to each of the 42 needs and 35 opportunities of the lists that represented combinations of DUs and contractor-identified technology requirements. The values of the scores were used to rank-order these lists, resulting in the *INEL Identified Technology Development Needs Prioritization*, (DOE, undated a) and the *INEL Identified Technology Development Opportunities Prioritization* (DOE, undated b).

STCG Contributions

These two products of the integrated product team—a ranked list of technical needs and a separate ranked list of alternative technology opportunities—were used by the STCG Working Group to explore possible funding sources for supporting the work, beginning with the EM office that represents the problem owner. For example, the highest-priority need—sodium-bonded spent nuclear fuel—has as a primary funding source the EM-60 problem owner. The OST program is a secondary funding source. This exercise accomplishes several things:

- If a satisfactory funding source is available apart from OST, then the need or opportunity is removed from the list that is eventually generated as STCG input into the OST program;
- The separate lists of needs and opportunities are combined into a single list because the OST program does not support this distinction (they are all representative of viable technology development R&D work).
- This list is then reprioritized, because the priorities of what to provide to OST are not derivable trivially from the previous work of the integrated product team. For example, a decision has to be made on the relative priority of an alternative technology *opportunity* versus a *need*, which were represented on two separate lists in the EM integration effort described above. The reference used for this reranking was the FY 1997 Murder Board Priority List (DOE, 1996k). The use of this reference restored priorities to those of the initial ranking of DUs. The entities that remain, however, are not the full set of the original 169 DUs, but rather those that remain after grouping of similar ones and after excluding those that have sufficient funding from non-OST sources.

These rank-ordered lists are then examined and adjusted by the upper-level STCG management (composed of both DOE and Lockheed Martin employees). These upper-level managers make the final ranking decisions based on their judgments, which, among other things, are assumed to incorporate local stakeholder inputs that have been solicited already. These judgments adjust the rank orderings in minor ways—only about 3 out of a total of 100 rankings were modified. The result is the final list of site technology development needs that are shared with Focus Areas, Crosscutting Programs, and Industry Programs, as input for their use in developing a portfolio of projects (Technical Task Plans, or TTPs) to fund.

HANFORD STCG

At a Hanford site visit in January 1997, the committee heard that technology development at Hanford involves project- and result-focused efforts, incorporating long-term science needs as well as technology needs. These efforts engage not only the STCG, but also the Tanks Focus Area, the Hanford Strategic Plan, and *Accelerating Cleanup: Paths to Closure* (DOE, 1998a), which are important programs with which the STCG interacts.

STCG Membership, Mission, and Function

Unlike Savannah River, where voting members are federal employees only, the Hanford STCG has voting members among contractors, regulators, stakeholders, and Native American tribes, although DOE ownership of the decisions is acknowledged. The STCG is a "corporate voice" in which these parties discuss their position on each topic considered. Their avowed mission is to

- identify and prioritize site problems and needs,
- assess and recommend technologies for potential use at Hanford,
- facilitate their demonstration and implementation at Hanford, and
- promote privatization and commercialization.

The STCG has a management council, to which four subgroups—one for each OST Focus Area—report. The management council is the final decision-making body for STCG outputs.

STCG Interactions with Other OST Program Offices

The STCG's major interaction with each Focus Area is in delivering a list of prioritized site technology needs. In the future, the STCG will demand quality answers from each Focus Area on its response(s) to each need.

The Hanford Tank Initiative was endorsed by the STCG's management council; one day after a Hanford STCG letter of endorsement went to OST Deputy Assistant Secretary Clyde Frank, $7 million of OST money was released for the HTI. The HTI is designed to develop technical retrieval information to inform future tank closure decisions and to support the privatization initiative.

STCG Activities of Service to the Site

On-site, the STCG formulates specific technology development recommendations in reports to local site management. An STCG member stated that the Washington State Department of Ecology was a partner (rather than taking a less active role), as the regulatory member of the STCG and as a signatory member in its reports. The specific technology development recommendations are based on three criteria: a site need for it, a technically sound offer, and stakeholder concurrence in it. Out of 12 to 15 such proposals to the STCG, 10 were endorsed, and 6 or 7 were funded.

The STCG also issues policy recommendations relevant to the implementation of new technology. Such recommendations have no regulatory weight but attempt to capture consensus among members. As an example, the STCG supported an *in situ* treatment zone remediation approach at the North Springs area, which was not implemented in part due to negative public feedback.

As an additional activity, the STCG championed the creation of the Hanford Technology Deployment Center (HTDC), a "portal" of outreach to private industry that is designed to facilitate the demonstration and deployment of private-sector technologies on-site.

STCG Prioritization and Decision-Making Processes

Through interviews with (contractor) project engineers, technology needs get fed into the STCG process. The management and integration (M&I) contractor determines the technical approach to a remediation job, from which are derived the problems that go to the STCG as technology needs.

The STCG records these needs with the following uniform protocol:

1. The statement of the site technical *problem* is cast to identify the *functional need*.

2. Out of this problem statement is derived the statement of the *technology need* that specifies the functional requirement(s) that would solve the problem.

The following example illustrates this approach:

1. *Sample problem*. Potentially dangerous mixtures of flammable gases in tanks. Here, the functional need is to prevent a serious combustion event.

2. *Derived technology need*. Some method to agitate waste to prevent the buildup of flammable gases.

Each subgroup (e.g., the Tanks Subgroup) has amassed a list of technology needs using this uniform nomenclature. At the June 1996 national STCG meeting, a common standard format was adopted by all STCGs, to provide complex-wide uniformity of needs statements.

The STCG management council has established a set of criteria, such as cost reduction, schedule reduction, and worker safety enhancement, against which needs can be scored (with a simple scale of 0, 1, or 2 corresponding to low, intermediate, or high, respectively). Each STCG subgroup (e.g., Tanks) evaluates each need against the list of criteria using discussion among its members to derive a numerical score in each category. The criteria serve as discussion points for each technology need under consideration. The sum of these individual scores is a total numerical score used to rank the needs. The result—a prioritized list of technology needs—is presented to the STCG Management Council for official STCG approval and to each Focus Area.

This STCG Subgroup process gets people involved by fostering discussion, the output of which is the determination of the scores (0, 1, or 2) of each need against each criterion. This people-based "massaging" approach is preferred by site managers over a more rigorous, quantitative analysis-type methodology.

The Focus Area is charged with developing TTPs, ostensibly to meet the technology needs of the complex. In the future, each Focus Area will be queried as to its response for each need identified by an STCG. Such responses might take the form of "this TTP was established to meet that need," or "site X has something for that" or "the commercial sector, specifically company X at location Y, has something for that."

STCG Tanks Subgroup

The Tanks Subgroup is one of the STCG subgroups that performs the above-mentioned prioritization exercise for needs related to tank waste in the DOE-EM complex. Additionally, this subgroup contributed input on the Project Hanford Management Contract (PHMC). This contract specifies the Fluor Daniel Hanford performance measures, which detail incentives to using innovative technologies to meet site needs and schedule commitments.

Attributes of Hanford STCG Process

During the committee site visit with STCG representatives, the following positive attributes about the STCG were mentioned by those in attendance:

- The process is systematic; any technology can be traced back to a site problem.
- Good discussions are held.
- Functional performance requirements are included in the description of a need.
- The process is oriented to have an outcome (i.e., there is an outcome, not just a process).

- The STCG provides a forum for early stakeholder involvement (greater acceptability is engendered if such involvement occurs in an early technology development phase, rather than at the end point of the process).
    - The STCG is an effective way to reach out to the public and to industry.
    - The STCG output is a real list of site needs and business opportunities, not just a "wish list."

Additionally, the following attributes, more negative in character, were mentioned:

- The ranking criteria are sometimes duplicative and overlapping (This implies that the criteria definitions and the weighting system could be made more rigorous. However, the criteria are merely tools, and are not sophisticated enough to use apart from some rationalization).
- Input data are imperfect, so the true underlying technical problem is sometimes not stated explicitly. This relates to how the needs are derived.
- The numerical outputs of the weighted scoring exercises are not completely trustworthy because of imperfections in both the input data and the weighting scheme.

COMMITTEE OBSERVATIONS

The STCG is a facilitator for information processing, in that it feeds information to the "real" OST decision-making bodies that allocate funds—the Focus Areas and Crosscutting Programs.

For technology development funding, two processes exist at the Hanford site. One process involves the user organization (e.g., EM-30) only, in which funds are allocated for a particular project, without involving the OST STCGs and Focus Areas. A second process goes through these OST program units, with all the attendant efforts placed on the prioritization and ranking of needs, identification of multiple site opportunities for deployment, and competition for limited funds.

In response to a question of what the biggest problem was, the DOE-Richland and STCG representatives in attendance at the January 1997 site visit gave an answer that new technologies still do not get deployed on-site and that budget cuts are not friendly to technology development, which is perceived as an add-on to the base program scope.

Another issue is whether OST should do "backup" technologies that result in improvements in risk, safety, and cleanup results, rather than fund what may be technically weak parts of the existing EM-30 and EM-40 baselines. That is, could baseline improvements be separated from off-baseline approaches, perhaps with EM-50 money going to the latter (for gains in the quality of environmental cleanup achieved) rather than the former (for gains in cost and schedule of already planned work)?[2] One related issue is when one stops perfecting a baseline (which is presumed to work or presumed to be EM-30's or EM-40's job to make it work) and directs technology development money toward off-baseline opportunities. Most of OST's current program is not directed at techniques outside the current flowsheets. This kind of programmatic redirection would be a major decision that is one step away from the current practice (i.e., currently the EM-50 program assists EM-30 and EM-40 flowsheets, and hence DOE and DOE contractor efforts at implementing them). Among other considerations, baselines are to be consistent with the regulatory Records of Decision (RODs), thereby creating a significant disincentive to change them and to work on different approaches.

In the ranking of technology proposals and needs, the current list of criteria is not rigorously weighted. One possible change would be to establish a hierarchy of criteria, to avoid double counting and better focus the discussions. One important criterion absent from lists shared with the committee is the probability of technical success—the probability that a particular technology would solve the site problem

---

[2]The INEEL site prioritization approach segregated baseline needs from off-baseline opportunities. The INEEL STCG later combined these two entities.

under consideration (and a related criterion, the necessity of achieving success on this problem). Because some of the criteria are screening in nature, multiplying scores, rather than adding them (as was the FY 1997 practice), might be a more appropriate weighting method.

In the site's process of identifying needs and technologies, it was not clear to the committee how non-DOE input (e.g., state-of-the-art private industry techniques, similar problems and solutions related to the National Aeronautics and Space Administration (NASA), or ideas from other outside sources) were incorporated into the development of the portfolio of ideas (i.e., the generation of site needs and their technical solutions).

Of all the major sites that the committee visited, the Idaho STCG employed the most rigorous methodology in generating a prioritized list of technical deficiencies. The numerical outputs from the multi-attribute utility analysis were traceable and readily amenable to updates, if a modification such as a change in a weighting factor was ever contemplated at a later date.

The process consumed significant (human) resources, which could be a potential problem in years of budgetary cutbacks.

By virtue of including contractors in the STCG composition, the priority site needs are reasonably well defined technically and are endorsed by management in charge of site operations.

Appendix C

Focus Areas

In the OST Action Plan of 1994 (DOE, 1994a), national Focus Areas were established to integrate technology development activities at DOE-EM sites complex-wide in order to prioritize efforts to address the greatest technology needs and to benefit from the results of program efforts at individual sites. By 1996, OST had divided problems into four major Focus Area categories: Subsurface Contaminants, Mixed Waste, High-Level Waste in Tanks, and Decontamination and Decommissioning. Areas of technology development applicable over a number of Focus Areas are implemented through Crosscutting Programs that until 1997 were managed through DOE headquarters (DOE-HQ).

A complex-wide inventory of problems and technology needs was conducted from 1994 to 1995 to provide information for decision making on a national basis. These needs were compiled and reported in *National Technology Needs Assessment* reports, which documented the prevalence of problem types and identified generally similar ranges of technology needs across the DOE complex.

This national inventory of needs formed the technical basis for the initial design of the Focus Area programs. The problem statements were analyzed by the Focus Areas for commonality of technical issues, an effort that also resulted in the transfer of some problem sets between Focus Areas. The resulting problem sets in each Focus Area were further grouped into distinct needs statements by combining similar projects to form work packages, which were prioritized in a technology needs portfolio. After preparation of a draft budget for the prioritized technology needs portfolio, Focus Areas representatives met with the STCGs, Technical Program Officers, and other stakeholders to present the prioritized needs programs and obtain feedback prior to presentation of the final program to DOE headquarters.

SUBSURFACE CONTAMINANTS FOCUS AREA

The Subsurface Contaminants Focus Area (also commonly referred to as "Subcon") is situated at the Savannah River Site, one of many DOE-EM sites with subsurface contamination by radionuclides, toxic metals, and organic compounds. The SCFA combines into one national Focus Area two former Focus Areas—Contaminant Plumes and Landfills—which were separately managed until 1996. To address site needs and DOE program priorities to identify and develop environmental technologies to address soil and ground water remediation problems, the SCFA has identified the following four strategic goals:

1. ability to contain and/or stabilize contamination sources that pose an imminent threat to surface and ground waters;
2. ability to delineate dense nonaqueous phase liquid (DNAPL) contamination in the subsurface and to remediate DNAPL-contaminated soils and ground water;

3. ability to remove a full range of metal and radionuclide contamination in soils and ground water; and

4. ability to remediate landfills that pose a continuing threat to surface and ground waters.

To meet these goals, the SCFA funds the development and deployment of innovative remediation technologies that have the potential to significantly reduce site cleanup costs, reduce risk, or provide benefits beyond those provided by the baseline technology. These technology development activities are conducted in accordance with *Accelerating Cleanup: Paths to Closure* (DOE, 1998a,). The SCFA obtains information on problems and technology needs from the STCGs using a standardized needs template (developed by the STCGs). Additionally, various activities are conducted by the SCFA to support DOE-HQ or interdepartmental initiatives.

Status in December 1996

In 1996, the national needs inventory resulted in more than 520 needs statements, which were consolidated into approximately 34 "problem sets" and approximately half that number of "work packages" (Table C.1) based on similarity of problems and needs. Subsequent solicitations are made annually to the STCGs regarding new or updated problems and technology needs. As of December 1996, most of the portfolio of SCFA projects were ongoing, late-stage projects, initiated prior to the development of the prioritization criteria process. The suite of SCFA projects in December 1996 primarily addressed treatment of contamination by volatile organic compounds. The SCFA intends to obtain complete engineering cost and performance data for 11 of the existing portfolio technologies by 1999 so that they can be moved into the private sector. The SCFA decided to continue funding for these more "mature" projects to completion within two to three years, after which the portfolio is anticipated to change to better represent national priorities.

Process Description

Innovative technology need projects have been identified and evaluated using recently developed prioritization criteria and decision-making procedures.

From Aggregated Needs to Prioritized Work Packages

The Focus Area work begins with the definition and validation of specific remediation needs at the sites. Problem or needs statements provided by the STCGs are reviewed and prioritized by the Focus Area into a selected number of consolidated work packages for eventual funding depending on budget availability. Actions made by the SCFA include the designation of specific lead staff positions responsible for gathering site needs statements, validating the needs statements, and developing a needs database that matches needs to technologies and identifies technology gaps. The Stakeholder Coordinator (SHC) is responsible for gathering and validating site needs through interaction with STCGs and other stakeholders. The Technical Team is responsible for reviewing and evaluating needs statements to determine whether the needs are within the mission of the SCFA and for identifying possible available technologies. The Technical Team is comprised of representatives of DOE personnel from SRS, INEEL, the Richland (RL) field office, and the Albuquerque (AL) field office. The Systems Engineering Lead is responsible for matching needs to technologies and identifying technology gaps through the use of a computerized decision support system.

TABLE C.1. The FY 1997 Work Packages of Subsurface Contamination Focus Area, listed in Priority Order

Designation	Description
WP2	Portable Selective Hot Spot Removal System Demo
WP3	In Situ Stabilization for Contamination or Removal
WP4	DNAPL Characterization (transferred to CMST)
WP5	In Situ Destructive Treatment Technologies for DNAPL's
WP6	Advanced Subsurface Containment Systems Design and Performance Validation
WP7	Mobilization, Extraction, Removal of Metals and Radioactive Contaminants
WP8	Long-Term Containment Systems Monitoring and Maintenance
WP9	Soil Removal, Segregation, and Treatment of Waste Unit
WP10	Secondary Waste Treatment of Extracted Ground Water
WP11	Reaction Zone Barrier Systems for Metals and Radioactive Contaminants
WP12	Innovative Alternative Containment System Deployment
WP13	Reaction Zone Barrier Systems for DNAPL's
WP14	Mobilization, Extraction, Removal Technologies for DNAPL's
WP15	In Situ Bulk Waste Treatment

NOTE: Not shown is WP1, program management, which comprises approximately 5% of the program budget. These work packages were used by program managers to build the program and budget to be aligned with user needs.

SOURCE: Adapted from Wright, 1996.

The process of identifying needs begins with the STCGs' defining needs from a technical standpoint and assessing these needs in the context of *Accelerating Cleanup: Paths to Closure* (DOE, 1998a). The problem or needs statements are submitted to the SCFA, which evaluates, prioritizes, and integrates the needs into an overall technology development program.

When the needs statements are received, the SCFA evaluates whether technologies are available within DOE or commercially to address any such needs. If technologies are commercially available, the Technical Team contacts vendors to discuss performance requirements and, if appropriate, puts the vendor in contact with the applicable STCG.

Following review by the SCFA, needs typically are segregated into five categories:

1. needs that satisfy 2006 Plan requirements and have a high end user commitment, warranting continued SCFA technology deployment;

2. needs that do not justify additional action either because the baseline technology is adequate or because the potential risk of developing a new, more economical technology is excessive;

3. needs that currently have commercially available technology solutions;

4. needs being addressed or that should be addressed by another program (e.g., another Focus Area or Crosscutting Program); and

5. needs that are currently not high enough in *Accelerating Cleanup: Paths to Closure* (DOE, 1998a) priority to warrant additional action.

The SCFA Technical Team evaluates the needs and combines them into work packages based on similarity of purpose and characteristics. A ranking and rating process, described below in more detail, is used to prioritize the list of technology needs in preparing final work packages communicated to the site STCGs through the Stakeholder Coordination Manager. The resulting prioritized list of work packages (Table C.1) forms the basis for soliciting technology development proposals.

Input From Non-DOE Sources

A subcontractor (Scientech, Inc.) is used by SCFA to evaluate the availability of existing technologies by searching public databases for available technologies to address site problems and needs.

Communication Plans

The SCFA communicates the results of the needs evaluation to the STCGs and stakeholders. Future plans for the SCFA include posting the needs template on the Internet to improve communications with the field offices and provide information regarding performance requirements and program processes. Other plans include increased industry participation to identify technology gaps and more private-sector involvement in the technology development and procurement process.

The Development of SCFA Prioritization Criteria

The STCG problem or needs statements form the basis of program design. These problems and needs are ranked using criteria developed by various groups at various times. The SCFA has employed a variety of criteria and weighting factors in its process of prioritizing technical needs and allocating budgets. This history is reviewed briefly here.

Initially, criteria based on DOE's Risk Data Sheets were used to establish Focus Area priorities. These criteria, selected by SCFA and DOE-HQ, were

- public safety and health,
- site personnel safety and health,
- environmental protection,
- mortgage reduction,
- pervasiveness of the problem,
- regulatory compliance,
- social or cultural impact and economic risk reduction, and
- mission impact.

A scoring methodology was developed to prioritize SCFA problems and technology needs using these criteria in March 1996. Subsequently, additional Strategic Investment Criteria were developed to evaluate whether a technology was within the mission and strategy of the SCFA before it was proposed for funding. These Strategic Investment Criteria were used to determine whether existing technology was adequate and whether the proposed technology development was consistent with SCFA goals and EM-50 policy. Additional questions were related to whether there was a customer committed to implementation, whether the basic science and technical or performance requirements were understood, and whether the proposed activity would meet the customer deadline and result in costs commensurate with benefits received.

These Strategic Investment Criteria were used in April and May 1996 to develop a draft FY 1997 plan and budget. This draft plan and budget were presented to stakeholders in May 1996 at a meeting attended by TPOs, stakeholders, Community Leaders Network (CLN) members, and SCFA members, which resulted in the development of modified criteria and weighting factors to be used in developing the final FY 1997 PEG. These modified criteria were

- technical credibility (30 points),
- cost reduction (18 points),
- advantage over baseline (10 points),
- public, personnel, and/or environmental risk reduction (10 points),
- deployability (10 points),
- regulatory or stakeholder acceptance (9 points),
- return on investment (8 points),
- user support (8 points),
- regulatory compliance (8 points),
- secondary waste reduction (4 points)
- applicability to multiple sites (3 points)
- not a duplicative effort (2 points), and
- industry or federal agency leveraging (2 points).

The process of applying these criteria begins with a screening evaluation of technical credibility to determine whether a work proposal is appropriate based on the approach, previous performance, personnel expertise, resources, and so forth. If the proposal does not pass this evaluation, it receives a score of zero. If it passes the evaluation, it receives a score of 30 multiplied by various criteria weighting factors.

Several assumptions are also included within these criteria. For example, the highest rated of these criteria is technical credibility, not risk reduction. This is because it is assumed that the baseline technology addresses the required reduction in risk, so if any new technology outperforms the baseline technology, it is assumed to achieve the necessary risk reduction.

Several issues were identified in using these priority-setting criteria to finalize the FY 1997 PEG. These included redundancies involving criteria such as regulatory acceptance and exclusion issues relating to cost and baseline technology criteria. As a result, SCFA solicited additional comments in an October 1996 stakeholder workshop. More than 22 issues were identified in the workshop, including the need for increased end user or stakeholder input in the scoring, more emphasis on STCG-defined high-risk sites, greater clarification of DOE-specific needs (i.e., higher ranking for radionuclides than other contaminants), the need for a method to assess the practicality of technology deployment, and the need to avoid duplication of regulatory criteria.

A revised prioritization process is reportedly being developed for the FY 1998 PEG and FY 1999 IRB. Additionally, the SCFA has indicated that the prioritization criteria will be revisited at critical points in the budgeting and technology selection process.

Portfolio Management Decision Processes

Management of the technology development program includes regular review and evaluation of technology selection. Once the portfolio of technology development activities is developed, SCFA uses two major types of review as decision-making processes in the course of administering the technology development program. These are

- independent, external program and technology reviews (peer and gate reviews); and
- formal Focus Area reviews (project technical reviews).

Additionally, a variety of review procedures are used within EM-50 in association with program management, planning, budget authorization, and administration. This includes administration of the program management budget (5 percent of the total SCFA budget).

Peer and Gate Reviews

As of December 1996, peer reviews and gate reviews were conducted outside the Focus Area to provide independent technical input into the decision-making process (Bauer, 1996; Frolio, 1996; Heeb, 1996).

Peer reviews (ASME, 1997a) are conducted under contract to DOE by the American Society of Mechanical Engineers. ASME assembles peer review groups on an as-needed basis to resolve technical or stakeholder issues or to confirm that the technical approach selected is appropriate and prepare a report to be submitted to DOE-HQ and the Focus Area.

Gate reviews (DOE, 1996n), to evaluate the maturity of a technology to pass into more costly engineering and field development phases (stage 4: Engineering Development), were to be conducted and managed by DOE-OR as of December 1996. The gate review uses the following criteria:

- technology need;
- technical merit of system under development;
- cost of developing the technology;
- safety, environmental protection, and risk;
- stakeholder and regulatory acceptance; and
- commercial viability.

DOE-OR would prepare a report of opinions and recommendations that is submitted to the Focus Area and incorporated in the SCFA decision-making process.

Technical Project Reviews

For reviews at other technology development gates, SCFA performs a technical project review using the same general criteria as the gate review, but managed by the SCFA Lead Office. The stated goal of these reviews is to assess the technical status of a project and evaluate whether it should be continued. Reviewers are selected by the Technical Team Lead, Product Line Managers (PLMs), and Product Line Integrators and may include experts from both within and outside DOE. The Deployment Plan input from the Technical Team includes periodic reviews to evaluate compliance with federal and state agreements. The peer review team is to be comprised of technical personnel knowledgeable about the problem and technology, and the peer review report is submitted to the SCFA Lead Office.

Technical project reviews may be stand-alone reviews of a technology project or precursors or follow-ups of a gate or peer review. This is a primary analysis tool for making program design and management decisions in the Focus Area.

Responsibility for SCFA Decision Making

Decision making in the SCFA is conducted by various individuals and groups within the SCFA team. This team includes the SCFA Lead Office, SCFA technical and systems engineering teams, headquarters sponsors, PLMs and product line integrators, and technical support contractors.

Technical Analysis and Project Review Teams integrate input and provide recommendations to the Lead Office Manager (LOM), who has ultimate decision-making authority. The Technical Team Lead first recommends whether SCFA should develop the technology or whether the solution is available from private industry. The Technical Team also coordinates the program with associated funding agencies such as Crosscutting Programs, the EM Science Program, Industry Programs, and other federal agencies.

The LOM makes policy decisions regarding program design and development. The Program Execution Manager (PEM) develops action lists to implement these policy decisions into the Annual Performance and Deployment Plans. Program activities are tracked by the PEM and PLMs, and recommendations are made to the LOM who issues final budget authorization. Other specific LOM decision-making issues relate to congressionally mandated programs and to developing plans and performance specifications for activities involving industry participation.

Funding

Because of the heavily mortgaged nature (i.e., commitments to provide future funding to ongoing technology development projects) of the program, SCFA funding was not available in FY 1997 beyond that needed to continue ongoing projects through the development process. In anticipation of funding above the mortgage level for the FY 1998 program, SCFA solicited proposals to address needs in support of strategic goals that were not satisfied by the FY 1997 program.

Some funding is conducted by SCFA before prioritization of the work packages. This includes fixed program management costs and the costs to support activities under the jurisdiction of DOE-HQ, such as international interagency initiatives. Funding for these activities is authorized and directly allocated by DOE-HQ.

Committee Observations

The work packages are stated in general language (e.g., "In-Situ Destructive Treatment Technologies for DNAPLs") that is of only limited use in specifying what types of projects within that subject area should be funded. Hence, extra effort at refining prioritizations of general topical statements would be less constructive than the use of expert technical opinion to aid in the process of project development and selection.

MIXED WASTE FOCUS AREA

The Mixed Waste Focus Area (MWFA) is a national program with a scope covering the mixed waste technology needs at all DOE-EM sites. It is managed at INEEL.

The MWFA has developed a prioritized list of technology deficiencies to form a basis for the program's request for proposals to national laboratories and industry. These deficiencies are associated with "treatment trains" of large-scale engineering systems that would be needed to treat various categories of mixed waste. The process by which these technical deficiencies were identified is discussed briefly below, and is based on OST documents provided to the committee and on a site visit to Idaho Falls in November 1996. A fuller description of this process is found in Beitel (1996).

Structure and Composition

A key part of the MWFA is the role of the Waste Type Managers, who are contractors at various DOE-EM sites knowledgeable about their site's mixed waste streams and the technical issues associated with their management, characterization, storage, and treatment. Each Waste Type Manager is supported by various Waste Type Teams that provide contacts to site operations end user personnel and site personnel responsible for meeting regulatory requirements. These Waste Type Managers and Waste Type Teams provide important technical input (Conner and Connolly, 1996); however, DOE employees make the final decisions on any funding allocations.

The MWFA conducts outreach to stakeholder groups at a national level, such as the National Technical Workgroup, the Interstate Technology Regulatory Committee, and the CLN.

Process for Technology Development Decision Making

The major steps of the process by which the MWFA conducts its business are described in further detail below.

Inputs to the Process: Statements of Needs

The MWFA obtains technology development needs from the STCGs of the major DOE-EM sites and from the sites' program development efforts associated with *Accelerating Cleanup: Paths to Closure* (DOE, 1998a). Another key input was the collection of Site Treatment Plans.

Use of EM-30 Site Treatment Plans. The EM-30 Site Treatment Plans (STPs), developed from 1992 to 1995 as the DOE-EM response to the Federal Facility Compliance Act (FFCA), was a useful resource to the MWFA for interfacing with EM-30 program plans. The STP identifies each mixed waste stream and the proposed treatment facility or process that would treat the waste in accordance with RCRA requirements. The treated waste would then be ready for disposal in a suitable facility. These treatment options have been forged between EM-30 and its regulators; OST was not a partner to these agreements. The regulatory deadlines and program requirements for each mixed waste stream are shown in the Integrated Master Schedule (IMS), a MWFA product.

Systems Engineering Approach to Establishing a Technical Baseline

The MWFA diagrammed the STP treatment options and combined similar treatment processes to arrive at a manageable number (approximately 24) of "treatment trains," which are generic processing flowsheets depicting engineering systems, each of which treats a separate class of waste stream. The individual components of these treatment trains were examined to determine whether the technology needed to fulfill that function already existed as a full-scale demonstration or the equivalent, or whether further technology development work would be needed to provide a proof-of-concept test. The components in the latter category became technology deficiencies; 7 of the 24 treatment trains had no such deficiencies, and 17 had at least one deficiency.

Prioritizing Process Flowsheets. Four criteria were developed to prioritize these 17 treatment trains, also called process flowsheets.

1. Impact was measured by the volume of waste to be treated in this manner, the number of site customers needing this process, the number of affected waste streams in the mixed waste inventory, and a measure of the hazard of the waste.

2. Potential savings was measured by comparison with projected baseline cost.

3. Maturity of the flowsheet was measured by the number of technical deficiencies it contained and the estimated time it would take to remedy each (i.e., the time needed for a technology development project to result in a pilot-scale demonstration to resolve the deficiency).

4. DOE commitments were measured by the degree of commitment of the DOE problem owner (DOE line office management) to solving the problem. A remediation job for which legal orders or regulatory permits were already written received the highest weight, while internally planned DOE initiatives that lacked regulatory drivers received the lowest. This criterion was used to include considerations of regulatory milestones and other schedule requirements.

Each flowsheet was rated on a scale of 1 to 5 against these four criteria. These ratings were combined in a weighted manner using relative weights of 40, 15, 25, and 25 percent, respectively, applied to the above-mentioned criteria. The result of this exercise was a prioritized list of flowsheets.

Prioritizing Technical Deficiencies. Separate criteria were developed to prioritize the technical deficiencies, defined as those components of at least one flowsheet representing steps needing technology development work. The five criteria are the following:

1. impact, as measured by the number of flowsheets needing the step, the number of times it was needed in each flowsheet, and the severity of the hazard addressed by the step;

2. critical path deficiency, as measured by an estimate of the number of years lost while a job remained undone as a result of the deficiency's not being addressed;

3. maturity, equivalent to the stage it is at in the "stage-and-gate" model;

4. functional requirements, a measure of the degree to which the performance specifications were known (in a rough conceptual way verses a quantified, documented fashion); and

5. DOE commitments, a measure of the urgency of the job to the DOE-EM line office management responsible for the remedial action.

Each deficiency was rated on a scale of 1 to 5 against these five criteria. These ratings were combined in a weighted manner using relative weights assigned to each criterion. The result of this exercise was a prioritized list of technical deficiencies.

The Needs Matrix. The MWFA displays these two prioritization results together in a "needs matrix" (Figure C.1) showing both the flowsheets and the deficiencies. The columns are the rank-ordered process flowsheets, and the rows are the rank-ordered deficiencies. The entries of the matrix are simply "X" marks to show which deficiencies are integral component parts of which process flowsheets.

Technology Development Requirements Documents. Each need or deficiency is recorded as a separate Technology Development Requirements Document (TDRD) that specifies the technical requirements needed, stakeholder and regulatory inputs, and user (i.e., EM-30) schedule for treating mixed waste inventories. This schedule determines the window of opportunity for technology development. Eighteen such TDRDs exist on the MWFA homepage. Three examples of technology deficiency areas identified

Deficiency	Debris Thermal Treatment	Elemental Mercury	Comb. Organic Thermal	Wastewater Direct Stabilization	Debris Non-Thermal	Sludge Thermal Treatment	Debris Stabilization	Lab Packs Stabilization	Unique Waste. Comp Gases	Wastewater Thermal	Lab Packs Thermal Oxidation	Wastewater Non-Thermal	Lab Packs Chemical Oxidation	Sludge Thermal Desorption	Debris Thermal Desorption	Compatible Organic Non-Thermal	Sludge, Extract Oxidation
1 Hg Amalgamation		X				X				X							
2 Hg Stabilization	X	X	X			X				X		X				X	X
3 Initial Char. (NDE/NDA)	X	X	X		X	X	X							X	X	X	X
4 Hg Removal/Extraction		X			X	X	X			X				X	X	X	X
5 Desorption Efficiency												X		X	X	X	
6 Material Handling	X				X						X			X	X	X	
7 CN Destruction									X			X					
8 Sorting/Segregation Imp	X		X		X	X	X	X			X		X	X	X		
9 Radiation Partitioning	X														X		
10 Ash Stabilization	X		X			X				X	X						
11 Salt Stabil (Cl,NO3)	X		X	X	X	X		X		X	X	X	X	X	X	X	X
12 Heavy Metal Monitoring	X		X		X	X				X	X		X	X	X		
13 Alpha Monitoring	X		X		X	X				X	X			X			
14 Hg Monitoring	X	X	X			X				X	X			X	X		
15 VOC Monitoring	X		X			X				X	X			X	X	X	
16 Aq/Org N-Therm Destr.					X					X		X	X	X	X	X	
17 SC CO2 Extraction					X					X	X	X	X	X			X
18 Fission Product Removal	X			X													X
19 Waste Form Perf. U/TRU	X			X	X	X	X	X		X	X	X	X	X	X	X	X
20 Comparative Analysis					X						X	X	X			X	
21 Internal Drum Pressure					X							X					
22 Improved HEPAs	X		X			X				X	X			X	X		
23 Evaporator Design												X					
24 Hg Filter	X		X			X				X	X				X		
25 Nitrate Removal				X				X									
26 Molten Product Decant	X		X			X				X		X					
27 Refractory Performance	X					X					X	X					
28 Sludge Washing												X				X	
29 Container Integrity					X												
30 Trace Metal Removal												X					

FIGURE C.1 "Needs matrix" developed by the MWFA. Rows show a ranked list of 30 technology deficiencies. Columns represent a ranked list of 17 generic process flowsheets needed to treat all of the DOE-EM waste streams. Crosses indicate which deficiencies affect which process flowsheets. In reading down a column, the marks indicate, for a particular flowsheet, which constituent deficiencies need resolution by further technology development work. In reading across a row, the marks indicate, for a particular technical deficiency, which flowsheets use that technology as an integral step in the process. SOURCE: Conner and Connolly, 1996.

by the above procedure and recorded as TDRDs are mercury amalgamation (DOE, 1996n), mercury stabilization (DOE, 1996m), and chemical oxidation (DOE, 1996l).

These results form the MWFA Technical Baseline, useful for planning what type of technology development work to solicit and award contracts to, as described below.

Contracting for Technology Development Work

For each technical deficiency, the MWFA solicits additional information on such issues as technical performance requirements, stakeholder and regulatory issues, and availability of off-the-shelf technology. The latter is done in part via a Request for Information (RFI), which is published in *Commerce Business Daily*.

Program managers next decide whether a solicitation will go to national laboratories or to the private sector (i.e., industry or universities). This decision is currently made in part based on an estimate of the maturity (i.e., what stage it has reached in the stage-and-gate model) of the technology needed (with lower-stage work often designated for universities and higher-stage work designated for industry). No overt competition exists between the private sector and national laboratories; separate deficiencies are announced to each group. The work slated for national laboratories is keyed to their specific facilities.

Based on technical specifications to meet program requirements, a call for proposals is made. The proposals received in response to this call are evaluated by a review team. They first screen proposals based on the following five screening criteria:

1. consistency with MWFA scope,
2. consistency with MWFA strategy,
3. lack of commercial availability,
4. technical credibility, and
5. duplication (i.e., Whether the work is duplicative of other ongoing work).

Next, the review team scores each proposal quantitatively against each of the following five evaluation criteria:

1. priority of deficiencies addressed (25 percent);
2. technical effectiveness (25 percent);
3. implementability (25 percent);
4. environment, safety, and health (15 percent); and
5. regulatory or permitting (10 percent).

The relative weights shown in parentheses how the final weighted score is assigned to each proposal. Program managers use these scores as input information to inform their selection of which proposals to fund.

Monitoring Progress of Ongoing Projects

The PI of each project interacts with MWFA program managers who monitor progress.

The End Point of MWFA Efforts

The end of a technology development activity that is desired by MWFA Program Managers is a close-to-full-scale demonstration of a technology on real or surrogate waste. The goal, in a privatization context, is to provide the private sector with the technology necessary to bid on future privatized mixed waste cleanup jobs. The output of OST technology development work would be available to bidders as proven, demonstrated technology that can be built and implemented as one process component of any full-scale operation.

With this strategy, demonstrations at the end of the technology development work are done in accordance with written test plans, and the demonstration results are recorded in a Technical Performance Report (TPR).

Committee Observations

The process by which technical deficiencies are identified, ranked, and used as a basis for RFPs by the MWFA is the most rigorous of those practiced by all the Focus Areas.

The MWFA has assessed that most mixed waste technology needs are in advanced stages of engineering development (i.e., high stage and gate numbers) and therefore that the Focus Area might go out of business within seven years, based on the FY 1997 level of funding, if the projects funded within that time provide satisfactory results. This assessment is based on assumptions that may change. Many of these assumptions, such as

- the requirements of the Waste Isolation Pilot Plant waste acceptance criteria,
- the actual configuration of mixed waste facilities that will be constructed and operated (as compared with the planned operations of the STPs), and
- potential changes at local and national levels to mixed waste regulations,

have yet to be fully resolved.

DECONTAMINATION AND DECOMMISSIONING FOCUS AREA

The Decontamination and Decommissioning Focus Area is administered at the Federal Energy Technology Center in Morgantown, West Virginia. The following information comes from a site visit by the committee in May 1997 and from Focus Area publications.

The DDFA mission is to "develop, demonstrate, and facilitate implementation of systems to solve EM-40's and EM-60's identified needs for acceptable decontamination and decommissioning of DOE's radiologically contaminated surplus facilities" (Hart, 1997). The DOE-EM D&D-related tasks include the deactivation of approximately 7,000 contaminated buildings, the decommissioning of approximately 700 buildings, and the dispositioning of radioactively contaminated materials (including more than 600,000 tons of metal, 23 million cubic meters of concrete in contaminated buildings, and 400,000 tons of metal currently in scrap piles). The major drivers for some kind of action are the high safety and health risks associated with working in aged and contaminated facilities, and the high cost associated with facility deactivation, surveillance, and maintenance (i.e., high mortgage costs).

Structure and Composition

The DDFA is run by DOE program managers.

Decision Processes

Recent decisions within the DDFA have been associated with two activities: identifying and prioritizing D&D-related needs within the DOE-EM complex and conducting the LSDP. The LSDP represents more of a priority for the DDFA than individual technology development projects, which were funded only at a level of approximately $1 million in FY 1997. The methods used to gather needs and to initiate the LSDP are described in further detail below.

Needs Assessment Activities

In pre-Focus Area days (Focus Areas were formed in April 1994, replacing the former Integrated Demonstrations Program), D&D-related technical deficiencies were identified based on a workshop involving outside groups. The report of this workshop is no longer used.

The DDFA's management was moved from DOE-HQ to FETC in 1995, when the LSDP program was begun. In early 1996, DDFA personnel visited existing STCGs and site "end users" to collect prioritized site needs. To compare these needs, a short list of scoring criteria and their relative weighting factors were developed in a separate process involving interactions between the DDFA, STCGs, and the CLN.

The criteria and weighting factors (DOE, 1996aa) were the following:

- *public safety and health* (12 percent), which addresses the potential negative impact of the stated problem on safety and health of the public;
- *site personnel safety and health* (15 percent), which addresses potential negative impact of the problem on the safety and health of on-site personnel;
- *environmental protection* (10 percent), which addresses the potential of the problem to cause release of hazardous or radioactive materials on-site or off-site including the potential damage to environmental resource, habitats and populations;
- *compliance* (11 percent), which relates the problem to compliance with regulatory requirements, laws, court orders, binding agreements, DOE orders, and administrative notifications by a regulatory agency;
- *mission impact* (10 percent), which evaluates the problem relative to the existing DOE-EM mission;
- *mortgage reduction* (23 percent), which addresses the problem's impact on the D&D life-cycle cost;
- *social, cultural, economic* (7 percent), which addresses the problem's impact on social, cultural, or economic concerns in areas or regions surrounding the site, including potential impacts on stakeholder trust; and
- *pervasiveness of need* (13 percent), which addresses the problem in terms of magnitude (e.g., severity and volume of contamination, number of facilities affected) and applicability across the DOE complex.

National D&D Workshop. The site needs and the sets of criteria and weighting factors were used in a national priority-setting workshop (DOE, 1996aa) sponsored by the DDFA. Focus Area personnel aggregated the 102 site needs into 31 major areas. Each of approximately two dozen participants scored each of these 31 D&D needs against each of the criteria. The result was a prioritized list (Figure C.2) of 31 national D&D technology needs, a nonbinding informative input (to create awareness) to the LSDP Integrated Contractor Teams and the Crosscutting, Industry, and University Programs, for their

Priority Rank	Need Title	Weighted Average Score	Public Safety & Health	Site Personnel Safety & Health	Environmental Protection	Compliance	Mission Impact	Mortgage Reduction	Social, Cultural, Economic	Pervasiveness of Need
1	Dismantlement of Large/Complex Equipment & Structures	6.57	2.69	8.94	4.06	5.69	6.94	7.88	3.44	9.31
2	Decontamination of Contaminated Metal	6.17	3.00	6.50	3.56	5.38	6.75	8.56	3.25	8.31
3	Decontamination of Spent Fuel Storage Basin Liquids	6.14	6.75	6.25	7.50	6.56	5.50	6.00	6.44	4.63
4	Characterization of Contaminated Surfaces	6.14	2.44	7.75	3.53	4.81	6.13	7.75	4.25	9.06
5	Material Recycle	6.03	2.63	3.75	4.75	4.06	7.31	8.50	6.56	8.75
6	Decontamination of Contaminated Concrete	6.02	3.31	6.38	3.88	3.56	6.50	8.00	3.50	9.31
7	Characterization of Inaccessible Areas	5.72	2.00	6.00	3.56	4.75	5.69	8.06	3.06	8.69
8	Asbestos Treatment	5.69	3.81	6.88	4.75	4.44	5.50	6.00	3.73	8.56
9	Dismantlement of Process Equipment & Structural Materials	5.67	1.69	8.81	3.13	5.00	6.56	6.50	3.00	7.69
10	Decontamination of Large/Complex Equipment & Structures	5.53	2.38	5.75	2.75	3.31	6.69	7.88	3.69	8.13
11	Worker Protection Clothing	5.43	2.00	9.00	2.38	3.75	5.31	5.88	4.00	8.44
12	Improved Instrumentation for Monitoring Hazards	5.06	4.31	7.75	5.50	4.50	3.94	3.38	3.81	7.44
13	Fuel Element Handling	5.03	5.88	7.50	6.63	4.25	4.31	4.25	4.50	3.06
14	Characterization of Volumetrically Contaminated Concrete	4.99	3.38	5.19	3.38	4.00	5.44	6.06	2.56	7.38
15	Characterization of Buried Objects	4.76	4.50	4.44	5.31	4.81	4.00	4.13	4.69	6.69
16	Improved Roof Systems & Roof Maintenance Techniques	4.44	2.69	5.50	4.69	3.00	3.75	4.88	3.19	6.31
17	Treatment/Disposition of Liquid Wastes	4.39	2.31	5.19	4.63	3.69	4.25	5.00	2.56	5.88
18	Treatment of Radioactively Contaminated Pyrophoric & Flammable Materials	4.20	5.19	7.19	6.06	2.94	3.50	3.38	2.38	2.50
19	Decontamination and Recycle of Hg-Contaminated Materials	4.18	3.94	5.38	5.38	4.44	3.94	3.88	3.56	3.00
20	Waste Packaging	4.06	1.88	6.31	3.31	3.50	4.44	3.50	2.81	6.06
21	Decontamination of Lead	3.99	1.81	3.94	3.56	4.75	3.88	4.06	2.75	6.44
22	Dismantlement of Concrete Encased Piping	3.78	2.31	4.63	3.00	3.50	3.69	4.81	2.06	4.19
23	Raschig Ring Removal	3.73	1.94	7.94	3.38	3.88	3.88	3.88	1.94	1.44
24	Aerosol Capture & Control	3.55	1.94	6.50	4.25	2.25	3.19	2.75	1.75	5.00
25	Improved Exhaust Treatment Systems	3.53	2.69	4.06	4.19	2.75	3.00	3.88	1.81	4.56
26	Characterization Data Management	3.52	0.88	3.63	2.38	3.13	4.56	4.69	1.75	5.25
27	Remote Viewing of Operations in Contaminated Areas	3.18	1.00	5.13	1.06	1.69	3.38	5.13	1.44	3.19
28	Characterization & Decontamination of Construction Debris (High Explosives)	3.13	3.38	5.88	2.56	1.94	2.56	3.19	2.27	2.00
29	Decontamination of Graphite Reactor Components	3.11	1.50	6.81	2.53	2.44	3.00	3.13	1.56	2.38
30	Improved Fixatives	2.65	1.19	5.19	2.19	1.38	2.19	2.50	1.00	4.13
31	Characterization & Decontamination of Construction Debris (Chromium)	1.75	1.69	2.00	2.75	2.25	1.31	1.69	0.75	1.31
	Averages	4.58	2.81	6.00	3.89	3.75	4.55	5.13	3.03	5.78

FIGURE C.2 Prioritized Ranking of the 31 D&D needs by the National Needs Assessment Workshop. SOURCE: DOE, 1996aa.

consideration, to guide their selection of technologies. This prioritization exercise will be superseded by FY 1998 priorities derived from *Accelerating Cleanup: Paths to Closure* (DOE, 1998a).

Funding Technology Development Activities

The DDFA funds some specific PI-type technology development projects as well as a select handful of LSDPs.

Technology Development Projects. The DDFA is open to any technology showing a cost, schedule, or risk reduction, or anything that would do a job that cannot be done by current methods. DDFA program managers can steer technology developers to the right program to seek development funds. An outside developer would be directed to the Industry Program, which has a separate budget (because, legally, national laboratories cannot compete with private industry) and well-prescribed rules of interaction governing the solicitations for proposals and the type of feedback that can be given. An in-house DOE developer (e.g., someone at a national laboratory) would be steered to funds controlled by one of the Focus Areas.

Essentially all DDFA technologies are at the gate 5-6 level. The funds for specific projects comprise about 10 percent of the FY 1997 DDFA budget of $10 million (see Box C.1), with additional D&D-related technology development work represented by approximately $13 million to $14 million in the budgets of Industry Programs and Crosscutting Programs.

Large-Scale Demonstration Projects

The cornerstone of DDFA's technology development program is a series of LSDPs, which are conducted at unused nuclear facilities already slated for D&D (by EM-40 or 60) that serve as EM-50 demonstration test beds for new technologies. The intent of the LSDP is to demonstrate the potential advantages of innovative technologies over commercial, baseline approaches during D&D operations.

LSDPs were instituted to address two DDFA issues:

1. EM-30, 40, and 60 are not willing to risk using technology that has not been demonstrated at a large scale, and
2. the private sector is reluctant to take the risk of developing a product for DOE that may not be used or may only have a single use.

As a result, DDFA program priorities and expenditures have shifted emphasis from FY 1995, when

---

BOX C.1 D&D Technology Database Software

In 1997, OST, through FETC, funded the development of a database by a company called Nuclear Expertise, Inc. (NEXI). This NEXI concept, entitled "Integrated D&D Planning and Estimating Software Tools for Nuclear Utilities" was part of a contract with DOE to develop planning and cost-estimating tools for DOE facilities slated for decommissioning.

Several databases have been developed by a number of private industries, utilities, government agencies, laboratories, and other countries that are conducting decommissioning. A review of the NEXI product indicates that it provides little if anything beyond what is already in existence.

Although this specific example was not a high dollar expenditure, it is one example of how the OST decision process did not screen efforts already in existence prior to initiating a project.

100 percent of the approximately $10 million budget for DDFA was for technology development. In FY 1997, 90 percent of the DDFA funding was for LSDPs, with only 10 percent for individual technology development projects.

The DDFA selects the LSDP facilities from among site proposals on a competitive basis, with criteria to evaluate how well suited each structure is to demonstrating a suite of technologies relevant to complex-wide D&D challenges. The Focus Area sets up a team of several contractors to run the LSDP, allowing windows of time and opportunity in the schedule for cost and performance data to be collected on each innovative technology to be demonstrated. This Integrating Contractor Team (also called a "Strategic Alliance") selects which new technologies are demonstrated, choosing candidates from among EM-50-funded projects as well as from technologies outside the DOE system (i.e., the private-sector and abroad). The DDFA showcases the LSDP approach as the solution to the problem of how to bring private-sector companies and technologies into the DOE-EM complex while educating the private sector on how to do business with DOE. The cost of an LSDP is shared between EM-50 and EM-40 or EM-60.

The DDFA philosophy is that using EM-50 money to underwrite LSDPs is a way to

- provide a full-scale demonstration of a new technology to EM-40 and 60 federal employees and contractors,
- obtain side-by-side demonstrations of innovative technologies with baseline methods,
- reduce supplier risk and liability associated with the first-time use of new technologies, and
- introduce these new technologies to the community of contractors who will be bidding on future

work at other DOE sites and who need to know of the technologies in order to use them.

As a measure of success, 30 technologies had been demonstrated as of August 1996 in the first three LSDPs, and 11 of these have been retained for use to complete the D&D of that LSDP facility, having proven themselves—in the judgment of the contractor teams managing the jobs—superior to the baseline methods. The new technologies have not yet been applied to jobs outside the LSDP where they were first demonstrated (as of May 1997), perhaps because the first demonstrations were done relatively recently (in August 1996). The self-imposed strategic goal of the DDFA is to perform eight LSDPs in appropriate facilities by FY 2002 that would demonstrate the technical capability to handle 90 percent of DOE-EM's D&D problems, as calculated in some yet-to-be-determined way to account for problem needs and their technical solutions.

The first three LSDPs, selected by the DDFA in the first solicitation during the summer-fall of 1995, have slightly different arrangements for the Integrating Contractor Teams or Strategic Alliances. For the Chicago Pile 5 (CP-5; see Box C.2) test reactor at Argonne National Laboratory-East, the contractors entered into a legally arranged cooperative agreement. For the Fernald Plant 1 Uranium Handling Complex, a hybrid model was used in which the fixed-price site management and operations (M&O) contractor teamed with others (Babcock & Wilcox, Foster Wheeler, and others). For the safe storage of the Hanford C Production Reactor, the Bechtel Hanford prime site contractor led the job, with others operating as subcontractors. These LSDP management arrangements are an experiment to get improved technology implemented in the DOE-EM complex. The past practice—to have a test bed demonstration somewhere, followed by an EM-50 push to encourage use of the newly demonstrated technology—resulted in few implementations.

Two more LSDP facilities—Building 779 (a plutonium-processing lab replete with glove boxes) at Rocky Flats and Building K-27 at Oak Ridge—were selected in 1996 in a second competitive solicitation by the DDFA. These planned LSDPs were subsequently canceled (NRC, 1998a).

Selection of LSDP Facilities from Among Site Proposals. The competitive selection process by which LSDP facilities are chosen is handled like an official solicitation, not because this is required but because it is deemed to provide a suitably rigorous way to make a defensible and fair decision in a site

BOX C.2   The CP-5 Large-Scale Demonstration Project

The CP-5 demonstration focused on the decontamination and dismantlement of the CP-5 test reactor facility at Argonne National Laboratory.   The project was "an aggressive campaign to screen and evaluate potential technologies for demonstration in four problem areas: characterization, decontamination, dismantlement, and worker health and safety" (DOE, 1997c), integrating the technology demonstrations with the schedule of ongoing D&D work.   The CP-5 LSDP focused on technologies emphasizing characterization, worker protection, robotics and remote systems, concrete decontamination, and storage pool filtration.

The LSDP included six demonstration "sets" (DOE, 1996c).   The first set of demonstrations—the Validation Set—was intended to fine-tune the overall planning, execution, assessment, and reporting processes.   Set 5 relates to the dismantlement of the research reactor, which includes the demonstration of the Mobile Work System and a number of "innovative and commercially available end effectors and tools" (www.strategic-alliance.org/).

Twenty-five separate technologies were planned to be demonstrated before the project terminated in mid-1997.   The centerpiece of the demonstration was the dismantlement of the bioshield and reactor core.   Tools and equipment to be demonstrated included a mobile work system (Rosie), a Dual Arm Work Platform, and a Swing Reduced Crane Control system.   Other technologies demonstrated at the CP-5 LSDP included the Pipe Explorer™ characterization tool, the Mobile Automated Characterization System, an X-Ray Fluorescence Analyzer, Empore Membrane Filtration, Surface Contamination Monitor, Pipe Crawler, Gamma Cam, and Concrete Milling/VAC-PACR—all part of the strategic alliance objectives of FETC.   Data are provided from these activities to the U.S. Army Corps of Engineers for cost analysis (DOE, 1997c).

The Integrating Contracting Team was led by Duke Engineering Services.   Other members of the team include 3M, Commonwealth Edison, Duke Power, ICF Kaiser, Florida International University, and Argonne National Laboratory.   In addition to project management and integration, this entity is responsible for technology transfer.   The CP-5 LSDP was intended to examine potential application of qualified technologies to other DOE sites with similar needs or to private industry D&D projects such as Commonwealth Edison's Dresden 1.

Benefits of the LSDPs are that they

• achieved meaningful technology demonstrations that qualify for commercialization and/or wider application throughout the DOE complex;
• expedited deployment of D&D technologies required to meet specific customer needs while meeting OST established ROI guidelines,
• identified technology activities that should be reviewed for continuing DOE support; and
• introduced commercial business practices to technology deployment, thereby illustrating DOE's commitment to performance-based strategies and contracting reform.

However, many of the technologies and teamworking activities have already been demonstrated on other decommissioning projects, in both commercial and DOE sectors (e.g., the Shippingport Station Decommissioning Project, the University of California, Berkeley research reactor, Yankee Rowe, Shoreham, Fort St. Vrain, Tuxedo Park, and other sites in the United States and abroad).   A peer review panel (ASME, 1997a; Love, 1997) found that some of the technologies termed "innovative" are in fact fully developed (see also Box C.3).   The redemonstration of these technologies on DOE-EM D&D projects is a standard "on-line engineering" practice to adapt equipment to new applications.

competition. The DDFA issues a call for proposals from DOE site operations offices for candidate LSDP locations. The proposals received are evaluated as follows:

1. Each person on a Technical Advisory Committee independently reviews each proposal based on criteria such as the significance of the demonstration, its cost-saving potential, the potential for complex-wide application of any useful results, the existence of a variety of challenging technical problems (to maximize potential application elsewhere in DOE), the site commitment to funding the D&D project, and favorable management arrangements. This 7-to-10-person team is constituted of DOE managers from EM-30, 40, 50, and 60, plus representatives of the U.S. Army Corps of Engineers (the Corps was not included in the first solicitation in 1995). Each Technical Advisory Committee member provides written qualitative comments on each of the criteria, documenting the strengths and weaknesses of the proposal.

2. An Evaluation Team of three people develops consensus on a final ranking for each proposal, by

- assigning a numerical rating (weight) for each criterion,
- quantitatively scoring each proposal against the criteria, and
- averaging the three member scores.

3. A Selection Official receives these numerical results and issues a letter to make the award, but not before the final step.

4. The selection results are presented to the Focus Area Steering Committee (Deputy Assistant Secretaries of EM-30, 40, 50, and 60), for their review and approval.

In summary, proposals are prioritized using, first, qualitative comments by the Technical Advisory Committee, and later quantitative scores by the Evaluation Team. The approval of the Focus Area Steering Committee is obtained before the Selection Official makes the final selection. The decision reached is a team effort and a product of Focus Area management efforts in which many participate.

The following are evaluation criteria and their weighting factors that were developed (DOE, 1996f) from this process:

- significance of the demonstration (20 percent)—scale and scope, potential to reduce cost and risk over baseline in similar future projects, end state;
- readiness of demonstration (20 percent)—current status of characterization and decommissioning plans, D&D contractor under contract;
- site commitment (30 percent)—funding from other organizations, consideration of stakeholder concerns;
- project management (30 percent)—Integrating Contractor Team in place, with perceived strengths);
- program policy factors—no weighting factor appears to be assigned to this criterion; it is used by the Selection Official to consider needs such as distributing projects among a greater geographical area; optimizing the use of available funds; addressing federal, state, and local political sensitivities; diversifying the types of facilities hosting LSDPs; and considering projects that enhance existing or planned activities of DDFA, including collaboration with STCGs.

Selection of Individual Technologies to Be Demonstrated at an LSDP Facility. The Integrating Contractor Teams and Strategic Alliances (SAs) at existing LSDPs have their own processes to identify, screen, and select the technologies for demonstration, but in general each uses the following evaluation criteria (Hart, 1997) to decide how to allocate the funds they have been authorized by FETC to achieve the objectives stated in their proposal:

- technology maturity,
- application to facility needs,

- application to DOE complex needs,
- ability to adequately measure performance,[1]
- compatibility with baseline D&D schedule,
- demonstration cost,
- expected improvement over baseline,
- waste minimization, and
- technology provider participation.

The Integrating Contractor Team decides which technologies are to be demonstrated, at a typical cost of $100,000 to $300,000 per demonstration, and presents this selection as a proposal to two DOE employees (one from the DDFA and one from the DOE Operations Office for the LSDP site). The DOE employees exercise line-by-line veto power. One issue they consider is to ensure that there is no duplicative demonstration at another LSDP. The approved demonstrations are conducted. This process has varied from one LSDP to another, based on the DDFA's experience and the different legal arrangements that define the interaction between the Integrating Contractor Team and DOE managers. Each LSDP has a target number of demonstrations to be done, during the D&D job, and payment of EM-50 funds to the contractor teams is made in stages as individual technologies are demonstrated (according to pre-approved test plans), rather than in a lump sum.

The U.S. Army Corps of Engineers is used as an independent party to validate cost and performance data. The full-scale engineering cost and performance data generated from each individual technology demonstration are recorded in an Innovative Technology Summary Report (ITSR), also known as an EM-50 "green book."

Cost Sharing Among LSDP Participants. The full costs of performing technology demonstrations are shared by the DDFA (EM-50), vendors of innovative technologies, and the owner of the surplus facility (EM-40 or EM-60). The DDFA contributes several million dollars to each LSDP. In general, DDFA funds represent the incremental costs associated with demonstration of innovative technologies (i.e., the difference between the cost to demonstrate the new technology on a section of a facility and the cost to use a commercial baseline technology on the same section of the facility). Costs for the use of commercial baseline technologies are borne by the problem owner. The new technology vendor is expected to share the cost of the demonstration since successful demonstration will validate the product and provide a rapid avenue to commercialization. Funds may also be contributed from other sources such as the private sector, other DOE departments, and other federal or state agencies. Some of the technologies used in an LSDP can also be submitted for (ASTD) funding. Since the ASTD program is for deployment of already-demonstrated technologies, technologies successfully demonstrated at an LSDP are well positioned for ASTD funds.

Impact of Accelerating Cleanup: Paths to Closure *on the DDFA..* The current 2006 Plan calls for $2 billion to be spent on D&D in DOE-EM until 2006, with $18 billion spent after that date. The crafting of site 2006 Plans has generated more and better data on DOE-EM D&D jobs than the DDFA originally had. For example, it requires EM-40 and 60 to show the *schedule* of D&D operations. The level of detail is also sufficient for the private sector to make a rational decision on investment in going after these jobs. Such additional information (provided as part of 2006 Plan activities) has helped OST program managers (Hart, 1997) justify the need for each of the LSDPs in the second round. *Accelerating Cleanup: Paths to Closure* (DOE, 1998a) is also credited with helping to identify a cost and schedule estimate for each baseline technology.

---

[1]Demonstration performance indicators are identified (Hart, 1997) as worker radiation exposure, health and safety, cost or performance, schedule savings, training requirements, secondary waste generation, equipment mobilization or demobilization, and ease of equipment decontamination.

DDFA Bringing Technologies into the DOE-EM Complex. As described to the committee (Hart, 1997), the term "innovative" as applied to the DDFA might be best defined as "never been used before in the DOE complex"—meaning that a technology does not necessarily have to be new in the non-DOE world for it to be considered for DDFA funding. For the CP-5 Reactor LSDP, where 20 innovative technologies were planned to be demonstrated, 10 of these were mature full-scale hardware systems from OST-funded technology development projects and 10 came from outside DOE (i.e., they were already in use in the private sector and abroad, see Box C.3). Of these 20 projects, 14 had been demonstrated as of May 1997. The 20 were screened from a beginning list of approximately 100 alternative technologies.

D&D-Related Mortgage Reduction. The decontamination and decommissioning of unused buildings is not a compliance-driven activity in general; therefore, only a handful of projects are done each year. The DDFA sees technical innovation to cut D&D costs as an excellent opportunity to achieve mortgage reduction. The rationale is that, with so few D&D projects slated for the near future, opportunity exists to develop new technologies to impact the baseline costs of future jobs.

Committee Observations

The 1996 prioritization criteria are generally stated (e.g., public safety and health, environmental protection, and compliance), which has the disadvantage that they are subject to a wide range of possible interpretations. This can cause an evaluator to react to the term used, thereby introducing the evaluator's own value system judgments into the evaluation of technical need. The correction would be to provide expanded statements defining these criteria in more explicit terms.

BOX C.3. Characterization Hardware: The Pipe Explorer™ System

The Pipe Explorer™ device was an OST development that had been demonstrated at FUSRAP, (Adrian, Michigan); Albuquerque, New Mexico; and Grand Junction, Colorado (DOE, 1996q) by 1996, but which was redemonstrated in the CP-5 LSDP.

The Pipe Explorer™ device was developed by Science and Engineering Associates, Inc. under contract with FETC. It has been used to transport various characterizing sensors into piping systems that have been radiologically contaminated. The device can be deployed through constrictions in the pipe, around 90° bends, vertically up and down, and in slippery conditions.

The device is protected from contamination to eliminate cross-contamination and false readings by means of an airtight membrane, which is disposable. When the canister is pressurized, the membrane inverts and deploys inside the pipe. As the membrane is deployed within the pipe, the detector and its cabling are towed into the pipe inside the membrane. Measurements are taken from within the protective membrane. Once the measurements are complete the process is reversed to retrieve the characterization tools. Characterization sensors that had been demonstrated as of the 1996 report were gamma detectors, beta detectors, video cameras, and pipe locators. Alpha capability had been developed by that time, but had not yet been demonstrated.

The system is capable of deploying in pipes as small as 2 inches in diameter and up to 250 feet long for pipe of diameter 3 inches or greater.

The redemonstration of the Pipe Explorer™ device at CP-5, after its development and demonstration at three other DOE sites, is an indication of a problem within DOE-EM, which might be cast as a deficiency in obtaining the broad knowledge and adoption or utilization of innovative technology as it becomes known and demonstrated. If this function were conducted by another program office, technology development funds would not have to be used for this purpose.

The committee also notes that not all must be combined in the same weighted analysis. For example, the criterion of worker risk could be handled as a go or no-go judgment based on some safety threshold (e.g., whether the innovative technology has more worker risk than the baseline method), instead of in a weighted fashion with other criteria. As a second example, the criterion of mortgage reduction could be taken out (i.e., not weighted with the others, as was done in the 1996 effort) and evaluated separately by DOE program managers considering the policy angles, rather than being combined with the weighted judgments associated with the other criteria. This issue arises because the weight of this category may be different in 1998 than in 1996, and such a difference may affect the outcome. The concept of mortgage reduction is key to *Accelerating Cleanup: Paths to Disclosure (DOE, 1998a).*

According to the committee's perception, the descriptors of the 31 needs of the 1996 needs assessment exercise (Figure C.2) are too general to define the technical program.

If the program goal is to engender private sector participation in DOE-EM D&D operations (rather than to develop technology in deficiency areas), then such a mission should be formulated only after thorough research (and documentation) to ascertain the validity of the underlying assumption that this is what OST needs to do.

TANKS FOCUS AREA

This section discusses the decision-making approaches of the Tanks Focus Area (TFA), which develops technology to remediate tanks at four DOE-EM sites[2]: the SRS, INEEL, the Oak Ridge Reservation (ORR), and the Hanford site.

Structure and Composition

The TFA is run by DOE program managers, who work as a team with contractors, many at the Pacific Northwest National Laboratory (PNNL).

Process Description

Tanks Focus Area decision making involves four major steps: technology need identification, technology need prioritization, proposal solicitation, and proposal selection. Since its inception, the TFA has employed a well-defined approach to these tasks, an approach that continues to be modified and improved. The current overall TFA decision-making process is summarized in Figure C.3 (Frey and Brouns, 1997) and discussed below. This discussion will be in the context of decision making for an arbitrary fiscal year N.

Identification of Technology Needs

The TFA technology needs identification for year N begins in the first quarter of the prior fiscal year (FY N − 1) by sending each of the four relevant site STCGs a call for identification of prioritized technology development needs. The format for recording needs is a template originally created by the

---

[2]The tanks at the Nuclear Fuel Services Plant near Buffalo, New York; are not addressed by the TFA.

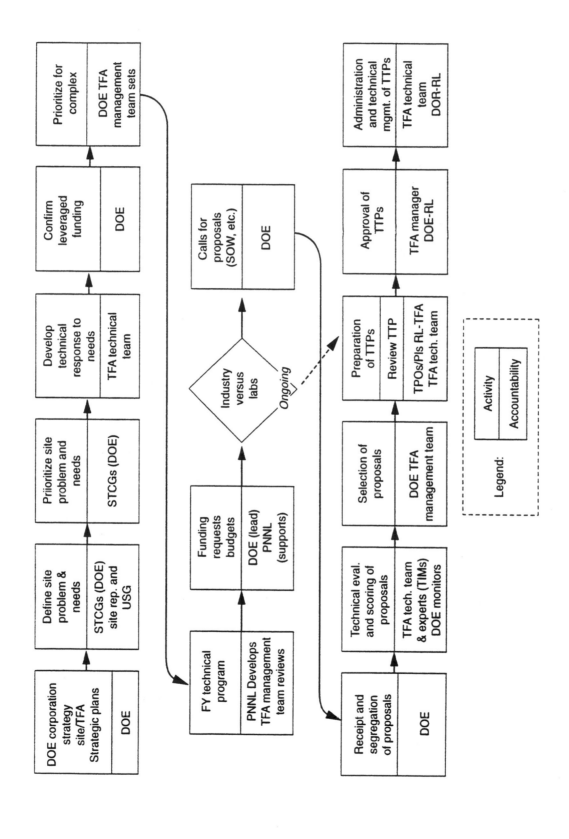

FIGURE C.3 Major Steps in the TFA process to select and administer technology development projects.
NOTE: SOW = Statement of Work; USG = User Steering Group. SOURCE: Frey and Brouns, 1997.

TFA. With the maturation and coordination of the STCGs, this template has been refined and standardized across all sites and Focus Areas. The call for technology needs results in considerable STCG activity to assess, establish, and prioritize site needs as described in Appendix B.

Aggregation and Prioritization of Technology Needs

The TFA sorts the prioritized responses from the STCGs by primary technical category (e.g., retrieval, characterization, pretreatment, and immobilization). The STCG priority valuations are retained. The needs are then analyzed to identify duplicate or similar needs (a common occurrence), which are aggregated into higher-level needs in a "roll-up" process. The priority of a higher-level need is assessed based on the STCG priorities of the constituent needs represented within the higher-level need category. This analysis and roll-up of needs is performed by the TFA Technical Integration Managers (TIMs), who are the second tier of the TFA technical organization and staffed by individuals from a number of DOE sites.

The aggregated needs are then subjected to a complex series of activities to prioritize them. First, the needs are prioritized by the TFA Technical Team. This draft prioritization is then reviewed by representatives of the four tank site STCGs and representatives of site contractor users via discussions and meetings. The result is a prioritized set of aggregated needs that is reviewed and approved by the various site STCG representatives. The prioritized needs are documented in an annual informal report containing the basic needs, aggregated needs, and final prioritization. This report is distributed to all affected sites. A prioritized list of site needs is achieved by negotiation among the four tank sites and the TFA technical staff using the criteria and trade-offs that each of these parties provides. A structured decision process and explicit criteria are not evident; in fact, the subjectivity appears to be desired by the participants. Thus, the participants appear to be satisfied with the process and the TFA performance continues to be highly regarded by other review groups and by DOE management (Berkey, 1998; Surles, 1997).

Creating a Work Plan of Proposed Projects

The prioritized, aggregated needs provide the basis for the TFA Technical Team to devise a technical work plan to meet these needs. The needs are apparently first assessed to determine whether they are carry-over needs that are already being addressed or can be addressed by ongoing projects. Ongoing projects appear to receive the highest priority; then new starts are considered. The Technical Team identifies projects to meet the needs, which includes ongoing projects.

Each need is evaluated to determine the way that projects to fulfill it will be solicited. Proposals may be sought from the private sector (industries and universities) or national laboratories by requests published in appropriate media, such as *Commerce Business Daily*, the *Federal Register*, Internet sites, and letters within the DOE complex. In some cases, a proposal is solicited from only one organization. These "sole-source" solicitations appear to be used for projects that require capabilities (e.g., facilities or access to wastes) unique to a specific organization. It is not clear to the committee how the Tanks Focus Area determination of unique capability is made and controlled.

The projects proposed (representing both the ongoing projects and the new starts), also called "technical responses," are prioritized with a quantitative procedure based on four criteria:

1. extent of multisite benefit,
2. cost or mortgage reduction,
3. support of TFA strategic goals, and
4. user commitment to deploy.

This prioritization process produces enthusiastic discussion, because the participants are normally aware of a nominal budget, which establishes a *de facto* cut line that provides motivation for proponents of specific proposals. The result is an initial prioritization that is documented and subjected to another review by the same group—representatives of the four tank site STCGs and representatives of site contractor users—that reviewed the aggregated list of prioritized site needs. The result is a prioritized list of TFA projects to be executed.

The results of this process are recorded during FY N − 1 in a formal document, a multiyear program plan, that outlines the midrange (several years) plans for the TFA. Additional information is provided during FY N − 1 to the DOE IRB, which is the planning basis for the fiscal year N + 1 budget, and to the PEG, which defines the specific work to be performed in the upcoming fiscal year N.

Process Evaluation and Committee Observations

As noted above, the TFA program is regarded relatively highly by other reviewers, which can be attributed in part to the strongly user-driven decision-making process. However, the process is complex and, in parts, opaque. Some of the specific concerns that it engenders include the following:

- The roll-up (aggregation) of site needs often renames and redefines the need of a specific site to the point that (1) the site does not recognize the aggregated need, even though the solution may be adequate; or (2) the original intent of a site need has been lost and the aggregated need does not meet this original need even though it is claimed to do so. Recent revisions of the needs template have attempted to address this problem by providing specific feedback to each site's stated need. However, it is not clear that the interaction between the TFA and the sites is sufficient to ensure that projects meet all the needs that are claimed.

- The means by which the TFA accommodates midcourse corrections necessitated by budget changes (apparently all too common) or changing needs at user sites (also common) are not clear to the committee.

- The TFA has been relatively successful in transitioning to a customer-focused technology development program. However, it is not clear that the TFA has effective mechanisms in place to terminate successful development projects and transfer emphasis to other areas. In some areas, investigators have developed yet-to-be-implemented technologies and are now proposing to pursue "second-generation" technologies while other aspects of tank remediation require more resources to develop first-generation technologies.

Appendix D

Crosscutting Programs

The OST Crosscutting Programs in the areas of (1) Robotics, (2) Efficient Separations and Processing (ESP), and (3) Characterization, Monitoring, and Sensor Technology (CMST), fund projects that in principle could be applied to the environmental problems of more than one technology or Focus Area. In the past, these programs were directed by managers located at headquarters. In FY 1997, the lead management for these programs was transferred from DOE headquarters to DOE field offices. This appendix summarizes information learned by the committee from DOE documents and from presentations by senior Crosscutting Program managers during a June 6, 1997, meeting.

ROBOTICS CROSSCUTTING PROGRAM

The Robotics Crosscutting Program is a national program, funding work by several PIs done at multiple DOE sites to support all Focus Areas. The program is based at the DOE Albuquerque Operations Office.

The Robotics Crosscutting Program uses a team of five to eight "coordinators"—contractor employees at several DOE labs with technical expertise in robotics—to interact with the Focus Areas and STCGs in their robotics needs and technologies. The committee heard (Yarbrough, 1997) that this team is responsible for ensuring that they do not develop what could otherwise be bought from outside DOE.

There is a chronology of events during a DOE fiscal year that leads to the selection of projects to fund—the important decisions—to define the next year's program. In crafting the program, the Crosscutting Program follows the lead of the Focus Areas (and STCGs) in dictating priorities. Based on these priorities, "initial guidance" of the robotics work to be carried out in the next year is determined and shared with the Focus Areas to solicit their feedback. "Final guidance" is then prepared that more fully defines the proposed scopes of work in each robotics Technical Task Plan (TTP). Part of the annual planning cycle includes gathering background materials and attendance at briefings, to gain knowledge of what the Focus Areas are planning. A key part of the Robotics Crosscutting Program planning for the next fiscal year is an annual planning meeting in May (a closed-door session with the coordinators), the result of which is a Major Thrust-Major Milestone micro TTP document (DOE, 1996t).

After receiving input from coordinators, a prioritized list of TTP projects to fund is developed, based on criteria such as

- user commitment (usually driven by a compliance deadline),
- probability of deployment,
- momentum (i.e., how far along the cleanup project is to date),
- currency (i.e., relevance),
- specificity (i.e., whether the need is defined well enough to be taken seriously), and

• clarity (i.e., how well organized the Focus Area is, as it affects the probability that the priority will not change in a year).

There is no written formula or recorded formalized process of ranking projects using these criteria. The coordinators can provide feedback to make minor adjustments.

The initial budget from OST upper management is a target value within a target range. Later in the fiscal year, the final budget figure for the following fiscal year is received. The Robotics Crosscutting Program budget is thus set in a top-down budget dictation by OST headquarters management. The FY 1997 budget was approximately $26 million, and the FY 1998 target budget is $11.7 million.

More than 50 percent of the Robotics Crosscutting Program budget goes to national laboratory projects; approximately 40 percent goes to industry; and approximately 10 to 15 percent goes to universities (Yarbrough, 1997).

Occasionally, project requests from EM-30 and EM-40 go directly to the Robotics program, outside the STCG and Focus Area routing system, and a few such projects are funded each year. These projects are cofunded, with joint financing between EM-30 or EM-40 funds and EM-50 Robotics Crosscutting Program funds.

The Light Duty Utility Arm (LDUA) was a robotics project funded out of the Tanks Focus Area, not the Robotics Crosscutting Program. This now-developed arm, initially intended for use in the Hanford tanks, was deployed there in September 1997 and is now scheduled for deployments at the Oak Ridge National Laboratory, the Fernald site, and the Savannah River site.

In his presentation to the committee, Dr. Yarbrough stated that he tried to "build systems that make sense," hoping that the two years needed to build a system do not result in a reorientation of program priorities and an associated lack of interest in using it. He estimated that one in three systems has been used (in a first-time deployment) to solve a real cleanup problem (e.g., an emergency contamination characterization problem for which a duct crawler was developed and used). His impression of the other two-thirds of the projects was that they involved equipment originally approved for construction that looked like a fit to priorities, but that the priorities and presumed applications did not materialize when the systems were finished two years later. Dr. Yarbrough stated that Focus Areas sometimes do not appear to know what they want well enough to provide significant feedback. He had similar difficulties with programs preceding the Focus Areas (i.e., the pre-1994 Integrated Demonstrations and Integrated Programs).

Development of a project was never stopped in midcourse. Performance measures for each TTP project are its own progress milestones and deliverables.

STCG needs documents are searched with key words to find robotics-related ones. Since essentially no money for new starts was available in FY 1997, those needs were compared to current projects to see what needs were being addressed by projects that were already funded.

The Robotics Crosscutting Program manager interacts with many groups, internal and external to DOE, primarily by giving presentations describing the program. The Robotics Crosscutting Program has multiple oversight by many interested outside review groups, interactions with both DOE and contractor personnel (in managing projects and assessing activities at DOE field offices), and interactions with multiple OST program units (i.e., STCGs, Focus Areas, and headquarters management). The lines of authority and communication within DOE are themselves a new construct, with management of this and other Crosscutting Programs at the field offices starting in FY 1997. The field-based OST program manager reports to his or her local line managers as well as to headquarters OST managers.

EFFICIENT SEPARATIONS AND PROCESSING CROSSCUTTING PROGRAM

The ESP Crosscutting Program is based at the DOE Oak Ridge Operations Office. A "Core Management Team" includes both DOE and contractor representatives.

The ESP Crosscutting Program has a mission to develop separation and treatment processes in partnership with the private sector and to scale up these results to achieve pilot-scale (or larger) deployments on DOE-EM wastes. The program has self-generated strategic goals. Important problem areas include

- tritium separation, needed by all STCGs;
- tank flowsheet uncertainties;
- transuranic separation, to reap the benefits of *not* sending waste to the Waste Isolation Pilot Plant (WIPP-bound TRU waste costs are estimated at approximately $30,000 per cubic foot, while on-site low-level waste disposal costs approximately $30 per cubic foot);
- mercury separation, if the ongoing Mixed Waste Focus Area efforts to procure a solution from industry do not succeed; and
- separation of cesium, strontium, and technetium from tank wastes.

Projects typically run one to three years, at approximately $250,000 to $400,000 per year. Based on annual reviews, some projects that show insufficient progress have been terminated. Others have been terminated if the needs change so that the separation technique under development is no longer required. One project was ended because of poor PI performance.

The ESP Crosscutting Program has three major processes:

1. a Review of Proposed New Tasks, discussed below in more detail,
2. a Midyear Review of Tasks, also discussed below in more detail, and
3. an Annual Technical Exchange Meeting (TEM), which is held in January, with published proceedings. This meeting, plus other inputs (such as "milestone reports," budget levels, interactions with program managers and field coordinators, and a favorable midyear review), determines whether ongoing projects are continued.

Proposed New Tasks and Their Review

The generation of new task ideas was described (Harness, 1997) as coming from national labortories' solicitations and from one industry solicitation through a PNNL request for proposal (RFP) a few years ago. Responses to these solicitations are reviewed by a group of typically four people—a member of the core management team, an academic technical expert, a former DOE (or DOE contractor) technical expert, and a representative from a relevant Focus Area (DOE, 1996d). The review criteria are

- Benefits (weight = 30 percent),
- Probability of success (weight = 20 percent),
- Implementation and task plan (weight = 20 percent),
- Qualifications of staff (weight = 10 percent),
- Resources (weight = 10 percent), and
- Cost of R&D (weight = 10 percent).

The numerical scores against these criteria are multiplied by the weighting factors, and the results are totaled. No cutoffs (i.e., threshold or screening criteria) are employed, nor are results multiplied (rather than added). The criteria and weights were developed (Mathur, 1997) in past discussions with core EM-50 management. This system has been continued to the present because it seems to work reasonably well.

Midyear Reviews

The Midyear Reviews are done in a two-step process described below. In the first step, statements of needs are used to draft RFPs. In the second step, these proposals are evaluated by an Independent Review Team.

Needs

As with all crosscutting programs, ESP Crosscutting Program priorities are in principle dictated by the Focus Areas. Hence, technology development needs come primarily from input from Focus Areas (which in turn obtain needs statements from STCGs), but additional needs come from three other sources:

1. extra STCG input independent of Focus Area interactions,
2. extra site input independent of a site's STCG, and
3. "experts."

In all cases (i.e., regardless of origin of the need statement(s)), the ESP Crosscutting Program interacts iteratively with Focus Areas to obtain their acceptance of the final short list of needs used as the basis for calls for proposals. The final needs from these Focus Area iterations are sufficiently mature to define performance goals that a PI would have to meet. The output of these interactions are performance specifications for calls for proposals. Proposals go out in the usual dual solicitation to national labs and the private sector (industries and universities), the latter via FETC Program Research and Development Announcements and Research Opportunity Announcements (ROAs).

Review of Proposals

A review and ranking of the responses to solicitations, which are PI-proposed new work tasks, is done by the following groups of people:

- ESP Crosscutting Program managers,
- Technical Review Teams of experts composed of academics and DOE contractor retirees, and
- Focus Area representatives.

The Midyear Reviews use the four criteria below.

1. *Benefit.* Specific considerations include the following:

   - Is a target problem identified, specific, and understood?
   - Is the corresponding baseline or best available technology understood?
   - Is the relative advantage of the new technology understood?

2. *Technical progress.* Specific considerations include the following:

   - Are partnerships appropriate and productive?
   - Have the correct technical issues been identified that must be resolved?
   - Are the technical accomplishments sound?

3. *Programmatic status.* Specific considerations include the following:

- Has there been progress toward DOE implementation?
- Has appropriate progress been made toward commercializing the technology?

4. *Future plans.* Specific considerations include the following:

- Is closure of the R&D effort defined, and are plans for the next fiscal year appropriate in this context?
- Is there a clear strategy for implementing the technology during and after closure?

Budget

The FY 1998 budget target is $5 million, down from the FY 1997 budget of $12.7 million. One-third to one-fourth of this money goes to industry; the rest is predominantly for national laboratory researchers.

The funds available for new starts are calculated by using an estimate of the program funding level for the next fiscal year. This estimate is provided by upper-level OST management, who handle interactions with external agencies (e.g., the Congressional Budget Office and Office of Management and Budget) to approve OST's overall budget and divide this budget among various OST programs.

Subtracted from this estimate is the "mortgage" (i.e., money tied up in ongoing projects) for the next year and the PRDA funding (the industry solicitation portion) associated with the program. The remaining funds are available for new starts in the national laboratory arena; however, at the current funding level, there are no new starts.

Program Accomplishments

Significant accomplishments credited to the ESP Crosscutting Program have included the following (Harness, 1997):

- crystalline silicotitanate (CST) ion exchange medium for cesium removal;
- the "High Temperature Vacuum Distillation Process" used to remove plutonium at Los Alamos National Laboratory (LANL), and transferred for a second deployment at Rocky Flats;
- the "Process Absorber Development Unit" of 3M-Empore, using its membrane technology for hosting selective ion exchange ligand media;
- liquid polymers, for mercury and RCRA metals (winner of an R&D 100 Award); and
- technetium speciation work, applicable to ground water at Oak Ridge National Laboratory (ORNL).

The ESP Crosscutting Program has sponsored work that has produced 40 peer-reviewed publications and 7 patents. No information was readily available (as of June 1997) on how many of the technology development projects led to deployments. Despite these accomplishments, the future budgets for this program are at $5 million or less for each year.

CHARACTERIZATION, MONITORING, AND SENSOR TECHNOLOGY CROSSCUTTING PROGRAM

Dr. Caroline Purdy, the HQ lead program manager of the CMST Crosscutting Program for 1991-1997, and John Jones, the recent field lead at the Nevada Operations Office, described the program during a committee meeting on June 6, 1997. This program has approximately 10 percent of its funds going to program support, twice the EM-50 target level, because of the personnel-intensive tasks of interfacing with people and marketing.

Decision Process to Fund Projects

The project selection process is briefly described as follows:

A. Technology deficiencies are identified by STCG and 2006 Plan documents. Note: Over time, there has been an EM-50-wide evolution of the quality and sources of information on needs (Purdy, 1997). The most recent needs statements are generated from planning exercises associated with *Accelerating Cleanup: Paths to Closure* (DOE, 1998a); previous needs references have been issued by the STCGs and the Focus Areas.

B. Calls for proposals are issued jointly by the CMST Crosscutting Program and the Focus Areas.

C. Proposals received in response to solicitations are reviewed during the CMST Crosscutting Program's annual review meeting in April by three groups of people. In this meeting, each PI performing ongoing work and those proposing new initiatives present their project in an open forum to an audience of technical reviewers and OST program managers.

The three types of reviewers at this meeting are the following:

1. *Technical peers.* Each PI's project is reviewed by three individuals from universities, industry, DOE, or other federal agencies, who score projects individually and then meet collectively (in a "closed-door" evaluation) to arrive at a consensus group recommendation. These reviewers judge each project against criteria (Purdy, 1997) that include

- technical quality (approach and merit);
- cost savings over baseline technology;
- past performance of the PI;
- collaborations;
- personnel qualifications, project organization, experience, and commitment; and
- facilities and equipment.

2. *User and FA representatives.* These are three individuals nominated by the FAs, who assess the relevance of the project to a user need.

3. *DOE program managers.* These individuals assess program policy factors and make the final decisions. Their considerations include budget and program balance. The typical project lasts three years, so in a flat-funding environment with perfect staggering of projects, one-third of each year's program money would be available for new starts. The achievement of program balance considers relative percentage of projects in support of each Focus Area and alignment with EM goals. The program policy factors are (Purdy, 1997)

- program drivers,
- safety and health risk reduction,
- regulatory and stakeholder acceptability,
- technology commercialization or implementation,
- breadth of application,
- projected future technology development needs, and
- balance between risk and reward, long term verses short term.

D. For funded work, project progress is monitored in three ways:

1. by facilitators, who are full-time DOE program managers, each of whom monitors from four to six projects and interfaces between the PI and the program;
2. through monthly technical progress reports; and
3. through annual technology summary sheets.

Budget and Future Demand for CMST Work Products

The FY 1997 budget was $13.4 million, and the FY 1998 target is $11.3 million in an era of declining budgets (the overall EM-50 budget was reduced by 25 percent in FY 1997 over FY 1996). One reason offered (Purdy, 1997) for the CMST Crosscutting Program budget's not being cut as much as that of other programs was that monitoring of the performance of subsurface barrier walls has been identified as the key to verification of their integrity, and EM strategy relying on barrier walls would therefore drive a significant investment in appropriate sensors and monitoring technologies. This strategy and rationale were employed to generate future budget targets (the DOE budget process has a two-year-out planning vehicle, as well as a one-year-out one), establishing in FY 1996 a high target for the CMST Crosscutting Program FY 1998 budget.

One technical need area discussed was for continuous emission monitors for offgas streams, an area that six CMST Crosscutting Program projects address, since these monitors are perceived to be crucial to the regulatory licensing and public approval of thermal (i.e., incinerator-based) treatment systems. Other major technology deficiency areas in high demand from many sites are

- long-term monitoring of subsurface contaminants and barrier walls;
- subsurface and D&D characterizations;
- in-tank sensors and sampling devices;
- process control monitors for high-level waste (HLW) treatment;
- mixed waste stream characterization to assist in a decision to treat versus dispose (presumably if one can show cheaply that one meets land disposal restrictions (LDRs), then one can dispose on-site via shallow land burial);
- D&D needs for nondestructive assay or examination of pipes and equipment;
- nondestructive assay or examination for RCRA metals and volatile organic compounds (VOCs) in WIPP-bound TRU waste; and
- moisture and corrosion detection for spent nuclear fuel (SNF).

Accelerating Cleanup: Paths to Closure (DOE, 1998a) cost calculations allow for a rough estimation of cost savings that could be achieved by new CMST-related technologies. These calculations require that a baseline method already be established; that a credible, more innovative method also be identified; and that good cost and performance data be available to make a comparison. A major goal for future starts is to impact cost savings using such projections.

CMST program managers have helped formulate ASTD proposals, and it is hoped that many of those chosen will be related to the CMST Crosscutting Program.

Information Resources and Technology Comparisons

Attention is paid to knowing the state of the art outside DOE. The resources known and used by the program include

- documentation of the status of sensor technology published by the Hazardous Waste Remedial Actions Program, and
- workshop reports by the EPA and individual national laboratories.

Comparisons of the performance of different technologies ("bake-offs") on the same test surrogate are sometimes conducted. In FY 1997, for example, such bake-offs were run (Purdy, 1997) for

- continuous emission monitors,
- non-destructive assay techniques (gauging performance on a sample 55-gallon drum of TRU waste),
- pipeline slurry monitors,
- cone penetrometer sensors, and
- wellhead VOC monitors (in a joint EPA-DOE verification test).

Interactions with Other Programs

The CMST crosscutting program interacts with the Industry Program (IP; to craft ROA and PRDA proposals based on CMST-related needs), the University Program (UP; to provide universities with CMST-related issues), and the TISA Domestic Program. Outside OST, the program has interacted with (Purdy, 1997)

- the Office of Energy Research, in its Small Business Innovation Research grant program;
- Disney Corporation, through a cooperative agreement not in the IP or UP;
- other federal agencies to develop cone penetrometer work via an interagency activity; and
- the Defense Advanced Research Projects Agency (DARPA).

Appendix E

Other OST Programs

Not all OST funds are allocated to the four Focus Areas and three Crosscutting Programs; in fact, only approximately 40 percent of the FY 1998 budget request was directed toward these seven programs. This appendix summarizes information learned by the committee about the other OST-funded programs and the decision-making processes used therein.

Two examples of these other OST programs are the Industry Program and the University Program. The IP and UP managers interact with Focus Area and Crosscutting Program managers to define relevant and appropriate scopes of work for the contracts, grants, and cooperative agreements that they manage. As with the Crosscutting Programs, the Industry and University Programs use the Focus Areas' prioritized lists of technology needs to determine work scopes. A process of communication among program managers leads to refinements of the concepts underlying the technical challenges that comprise the technical specifications of the solicitations. These two programs, and others, are described in more detail below. The following information comes from DOE publications, from a May 1997 committee site visit to the FETC facility in Morgantown, West Virginia, and from a November 1997 committee meeting in Washington, D.C.

INDUSTRY PROGRAM

The IP receives OST funds that are available for technology development work in the private sector (i.e., for PIs outside the DOE-EM complex and national laboratory systems). The major role of the IP is to translate needs provided by the Focus Area and Crosscutting Programs into technical performance requirements for solicitations to permit companies to submit bids. The IP strategy is to provide a site-independent, "brokering" mechanism between the OST program office requesting the work (and generating the technical specifications of the solicitation) and the private sector.

The IP is administered by program managers at FETC, a government-owned, government-operated contracting shop with expertise in fostering DOE partnerships with private industry. The IP issues solicitations to the private sector with two types of procurements: ROAs and PRDAs. Each of these announcements is for work related to Focus Area and Crosscutting Program technical topics. The IP funds are controlled separately from these of the Focus Areas and Crosscutting Programs; hence, program managers of those program units must coordinate with IP managers to arrive at an approximate amount of IP money and an approximate number of IP projects that will be funded each year on behalf of each Focus Area and Crosscutting Program.

The IP relies on the outputs of the Focus Area and Crosscutting Program units to generate its statements of technology development needs that are used to develop solicitations to the private sector. Iteration and feedback among program managers is used (Bedick, 1997), because the initial needs statements are sometimes not sufficiently well-defined to serve as technical specifications. The DOE procurement process is then followed, with contracts awarded as phased procurement vehicles so that

OST program management has the flexibility to terminate funding of an activity at specified phases of its development.

Another "iterative-collaborative" process is the decision of how much of the IP funds goes to support work in each of the Focus Areas and Crosscutting Programs. Equity issues arise in these negotiations, but the FETC office claims to have no bias or partiality since it is not a DOE-EM site with environmental cleanup needs. Targeted funding designations for the amount of IP funding to each Focus Area and Crosscutting Program are made in future DOE budget projections (i.e., IRB), but the final budget allocation depends on how well a particular solicitation succeeds (i.e., the industry response to the solicitations offered). The Source Selection Official (SSO) running a solicitation has significant say in this; formally, these decisions are made by the FETC director or designee, which in practice usually amounts to agreeing with the SSO.

The two major decision points of the Industry Program (Bedick, 1997) are:

1. the outyear program planning, done in collaboration with Focus Areas and Crosscutting Programs (i.e., budget targets for future solicitations and the specific technical subjects of future solicitations), and

2. "internal to FETC processes"—the iterations to define technical procurement specifications out of generally written statements of needs, as discussed above, which are collaborative with the same program units.

Since its inception in the early 1990s, the IP has reviewed 1,500 proposals and has expended $186 million in 83 separate contracts. Only about a dozen companies have been repeat winners of a procurement contract. The award of a contract typically takes about nine months. Of the 83 projects, 18 were terminated before completion; 18 were funded to completion (i.e., to the level of maturity represented by gates 5 or 6); and 47 are ongoing. Fourteen projects have already been deployed at multiple DOE sites, with others ready for deployment in about a year. The projected $1.5 billion cost savings associated with the 14 technologies already deployed, using some of the assumptions in the U.S. Army Corps of Engineers comparisons to baseline technologies (Bedick, 1997), implies about an 8:1 ROI for the IP.

FETC Procurement Process

The DOE procurement process is applicable to both PRDAs and ROAs solicitations of the IP. PRDAs are one-time solicitations based on specific technical specifications that are open for 45 to 90 days. ROAs are broad, general statements of need that are open to bids for a longer time (typically a year). The process detailed below was established to ensure compliance with Federal Acquisition Regulations (FARs), Department of Energy Acquisition Regulations (DEARs), and FETC standard operating procedures (SOPs) (Christy, 1997).

Proposals received in response to a solicitation are first reviewed by a team of four reviewers, consisting of an FETC project manager, a Focus Area or Crosscutting representative, a field representative, and an outside expert (from academia, industry, or a national laboratory). The first two are federal DOE employees; the last two need not be. The first and fourth are selected by FETC, the second and third by Focus Area or Crosscutting Program personnel. The four individuals review a proposal separately, noting strengths and weaknesses as judged against four criteria:

1. technical approach (weight 33 percent);
2. merit of the technology (weight 33 percent);
3. personnel qualifications, project organization, experience and commitment (weight 20 percent); and
4. facilities and equipment (weight 14 percent).

The four reviewers then meet to translate their individual assessments into a collective numerical score (on a scale from 1 to 5) against each of these criteria. These ratings, combined with the predetermined weighting factors associated with each of the four criteria, produce an overall numerical rating for the proposal.

This result is then transmitted to a Technical Evaluation Committee (TEC) of DOE employees, typically FETC and headquarters personnel, who consider additional "program policy factors" such as regulatory impacts, cost sharing, available funds, the degree to which the work would complement or enhance an existing effort (or, as a negative, whether it is duplicative of other work that would meet the same particular need), whether the approach meets the policy intent,[1] and whether the approach is truly novel and innovative. The TEC makes a recommendation to the SSO, who decides which proposals to fund. Usually, this official picks what the TEC recommends.

The solicitation process ends with a "quality debriefing" meeting to all parties that submitted proposals and that elect to attend. These debriefings give legally permissible feedback on general considerations for the way proposals were rated to assist proposal writers to produce better submittals in response to future solicitations. Sometimes a bidders' conference is held before a solicitation is ended, to respond to bidder questions and requests for supplemental technical information by giving this information to all bidders.

The integrity of this review process is enforced by FETC program managers. Government procurement restrictions dictate conflict of interest requirements on reviewers and the nondisclosure of proprietary company information.

Committee Observations on IP

The arguments (Markel, 1997) that are made about using FETC as a preferred procurement office for OST are that it has

- lower overhead costs than DOE field operations offices;
- no partiality (i.e., no conflicts of interest as to which DOE-EM site is favored in any decision, since FETC is not itself a DOE-EM cleanup site);
- expertise at soliciting from industry in a fair and open competition with a national (nonparochial) focus, in a way that is uniform because all Focus Area requests are treated the same way; and
- an established "one-stop shop" for industrial players who do not want the difficulty of negotiating procurements with separate DOE sites that have differing ground rules.

The committee observes that the purchasing practices of OST are at best duplicative, and appear to be confusing and cumbersome and to lead to poor purchasing decisions in some instances.

UNIVERSITY PROGRAM

The University Program is at present a collection of four groups of universities that receive funds through grants and cooperative agreements because of congressional appropriation's language. The OST program managers have limited power in directing the use of these funds. Each annual grant is approximately $3 to $5 million, and the total annual program is funded at approximately $18 to $19 million.

---

[1] As reported by FETC program managers, a major difficulty in reviewing technology development proposals lies in understanding the needs of the DOE-EM complex and the relevance of the proposed technology to meet these needs.

The UP participants include (Bedick, 1997)

- Florida International University, which through its Hemispheric Center for Environmental Technology conducts test and evaluation activities in support of the D&D and Tanks Focus Areas as well as the CMST Crosscutting Program;
- Florida State University, which through its Institute for Central and Eastern European Cooperative Environmental Research conducts joint technology development with Russian and Central European organizations, including a Czechowice oil refinery cleanup project;
- Xavier and Tulane Universities, which conduct biotic and abiotic studies on hazardous waste and novel modeling approaches for transport; and
- Mississippi State University, which through its Diagnostic Instrumentation and Analysis Laboratory supports several Focus Areas with specialization in instrumentation and control systems.

These universities are funded to do work that is of a more applied engineering nature than typical basic science university research. The University Program manager strategy is to make an effort to identify DOE-EM problems, cast them in appropriate terms, and offer them as research opportunities to these academic institutions. For example, one university conducted a thermodynamic modeling analysis in rheology, which is relevant to the DOE-EM problem of high-level waste in tanks and in the mixing that occurs during a melter pour.

TECHNOLOGY INTEGRATION SYSTEMS APPLICATION INTERNATIONAL PROGRAM

The TISA International Program of the has supported DOE-EM collaborations on technical projects done through cooperative agreements that have been signed by high-ranking officials (outside OST). So far, four such agreements have been signed.

One is with the Former Soviet Union (FSU), because of an administrative directive to stabilize the post-Cold War nuclear community and the fact that the FSU has similar cleanup problems, first-rate expertise, and a 1:30 labor cost ratio compared to the United States.

Another is with Poland (through the FSU agreement), which has similar contamination problems and a 1:10 labor cost ratio.

A third agreement is with Argentina. This agreement was signed for the following reasons:

- the Quadro-Pact Agreement of 1994,
- an emphasis on environmental cooperation at the December 1994 Summit of the Americas, and
- the Department of Commerce listing of Argentina as one of the top 10 emerging markets.

The fourth agreement is a DOE cooperative agreement signed with the Atomic Energy Authority (AEA) of the United Kingdom, primarily for application of AEA HLW tank technology in DOE-EM.

When asked by the committee about countries such as France and Japan with developed nuclear technology, OST program managers responded (Walker, 1997) that those countries with developed nuclear expertise were already represented in EM cleanups by the partnerships of companies such as COGEMA and Seimens with American companies in bids on EM cleanup contracts.

TISA DOMESTIC PROGRAM
TO FOSTER REGULATORY AND STAKEHOLDER ACCEPTANCE

The OST has long supported efforts to work with stakeholder and regulatory groups to obtain their support for the introduction of new technologies applied to EM cleanups. The decisions of what work to fund in this area are made largely within the TISA Domestic Program.

One example of the type of work supported is that of the National Technical Workgroup (NTW), an interstate group of regulators of mixed waste, who meet to share information on common issues that arise in their experiences of permitting mixed waste storage and treatment facilities and operations. However, not all stakeholder and regulatory interaction work is funded through the TISA Domestic Program. For example, the Subsurface Contaminants Focus Area works directly with the Southern States Energy Board (SSEB) in obtaining regulatory approval for new technology. The key idea in support of OST involvement with both the NTW and the SSEB is that each activity was designed to foster regulatory acceptance of a technology in other states if it were successfully demonstrated in one state.

Stakeholder interactions are handled in each Focus Area through interactions with the Community Leaders Network. Each major DOE-EM site has a Site-Specific Advisory Board (SSAB) or Citizen's Advisory Board that is funded by DOE-EM, as well as other interest groups supported by local government or private institutions.

OST investments in these activities show that attention to nontechnical barriers, such as stakeholder and regulatory acceptance of new technology up for demonstration, has been a program priority.

The TISA Domestic Program (formerly known as the Office of Technology Integration (OTI)) funds the STCGs, the gate and peer reviews, and a host of other activities designed to promote the awareness of OST efforts and to gain stakeholder and regulatory acceptance of new technologies.

This program fulfills its general mission to facilitate the acceptance of new technology by funding five types of work (Lankford, 1997):

1. decision integration, meaning cost savings analyses, stage-and-gate management, peer reviews, and ITSRs (also known as "green books");
2. technical application and implementation facilitation, including STCGs;
3. regulatory cooperation, including the interstate technology regulatory cooperation;[2]
4. public and tribal participation, including the CLN; and
5. information for decisions, including the Technology Management System and communications such as newsletters and web pages that promote awareness of OST programs and disseminate specific project accomplishments.

This program suffered a significant budget cut in FY 1998, which necessitated a priority ranking of work packages.

ENVIRONMENTAL MANAGEMENT SCIENCE PROGRAM (EMSP)

The EMSP funds basic research at early stages in the stage-and-gate maturity model. The EMSP has a congressional charter (U.S. Congress, 1995d) to consider basic research for novel remediation methods. The EMSP has a structure similar to that of the OST Focus Areas—while technology needs identified by the STCGs go to the Focus Areas, science needs identified by site *ad hoc* workshops go to the EMSP. Research proposals submitted to the EMSP for funding undergo two reviews—one by the DOE Office of Energy Research to assess the proposal's scientific merit and another by DOE-EM to assess the proposal's potential relevance to site cleanup problems (i.e., to ensure that the science needs it addresses are tied to some EM site functional cleanup requirements).

This national program collected science needs from the major DOE-EM sites. At a committee site visit to Hanford in January 1997, John LaFemina discussed this site's EMSP "Science Needs Process" resulting from a two-day Hanford workshop in 1996 sponsored by the Keystone Group, the Tanks Focus Area, and the Wiley Laboratory (also known as the Environmental Management Science Laboratory). In

---

[2] The TISA Domestic Program funded a similar program from 1992 to 1996, the Western Governors' Association Committee to Develop On-Site Innovative Technologies.

this exercise, workshop attendees identified 38 site science needs (23 of which were for tanks) and prioritized them by a list of criteria.

In principle, the EMSP science needs and successful science-based projects should transfer to the technology development program, perhaps by evolving to technology needs and/or technology development projects. The EMSP is at too early a stage in its development to have a substantial basis for judging its effectiveness and the nature of any of these kind of interactions.

Successful research results of EMSP projects should feed into OST's RD&D process. If this happens on a significant scale, EMSP decision making could substantially influence the overall cleanup of DOE-EM sites. In practice, however, the connection is tenuous. OST's technology management difficulties are not at the beginning of the pipeline. On the contrary, it is the pileup of projects awaiting either demonstration or deployment that has raised questions about OST's success. In addition, with the relatively small budgets for EMSP and the early stages of OST's RD&D process, the EMSP decision-making process is not likely to have much influence on the overall performance of OST over the next three to five years. A time horizon of 10 to 20 years is needed for a significant relationship to emerge.

Although EMSP decision making does not affect the near-term success of the OST technology development program, the EMSP decision-making process for proposal selection does appear to function satisfactorily. A previous NRC report, *Building an Effective Environmental Management Science Program: Final Assessment* (NRC, 1997a), reviewed the proposal selection process for FY 1996 and concluded that (1) meritorious projects appear to have been selected, (2) collaborative efforts were well represented, (3) new DOE researchers were attracted, and (4) where firsthand information was available, the committee was able to confirm the overall quality of the merit review panel.

ACCELERATED SITE TECHNOLOGY DEPLOYMENT PROGRAM

In 1997, with increased attention by the House Commerce Committee to the issue of a Return on Investment on the $2.6 billion expended since the start of the program in 1989, OST headquarters management created the ASTD program (formerly known as the Technology Development Initiative) as an EM-50 competitive solicitation to DOE field offices for their proposals of how they could use EM-50 funds to reduce a site's cleanup costs. The intention of the ASTD program was to award EM-50 money to a site to underwrite the first-time use of a new technical approach applied to a cleanup project already scheduled for the near term. A requirement of the proposals was to perform an ROI calculation showing the cost savings (or cost avoidance) relative to the baseline remediation method that would be used in the absence of EM-50 funds. This requirement implied that competitive proposals would go only to those cleanup jobs for which a baseline remediation method already existed and that a credible cost estimate existed for the baseline method.

The ASTD program was not designed to be a demonstration program for unproven technologies. The technical methods under comparison had to be proven techniques, with sufficient engineering and performance data to use as a basis for cost estimates. The better-than-baseline technologies could come from any source, either from outside DOE-EM or from formerly funded DOE-EM technology development projects that had matured into viable cleanup techniques. A subordinate aim of the program was to determine the degree to which technical innovations that enjoyed past funding support by EM-50 were represented in the suite of new and cost-saving technologies that were part of successful bids.

This program is an explicit attempt by EM-50 program management to address the barrier to implementing new technologies in DOE-EM site cleanups. It has been argued (Frank, 1997b) that, strictly speaking, since no new technology was under consideration for development in the ASTD, this program should not be funded by discretionary EM-50 funds. Instead, other DOE-EM offices should be the proper programs to fund the ASTD activities proposed and selected, activities that reported more than an order of magnitude in cost savings (or cost avoidance) to DOE-EM.

The ASTD effort to review site and field office proposals did not formally involve interaction with other OST program units. The initiative for generating proposals lay with the field office representatives of the community of problem owners and technology users (i.e., EM-30, 40, and 60). Although STCGs and Focus Areas were not formally involved, the network of site contacts made by the STCGs and the Focus Areas was most probably an asset to the proposal writers, particularly in finding other DOE-EM sites for second and third applications of a first-time remediation technology.

A proposal selection committee chose 20 proposals for funding out of 89 received (Walker, 1997b), using $27.7 million of FY 1998 funds. These 20 proposals were consolidated into 16 projects, each with a typical duration of two years. They represent collectively a ROI of greater than $1.3 billion, if each project is funded to completion by OST (this would require a continued commitment of OST funds of $21.1 million, $10.9 million, and $1.4 million in FY 1999, FY 2000, and FY 2001, respectively) and replicated at other sites, as specified in the terms of the proposals.

Appendix F

Industry's Best Practices in
R&D Decision Making

To make the best possible decision in a given situation, an organization's decision-making process can be examined to identify strengths and weaknesses using the decision quality framework shown below. This six-element framework is applicable to all types of decisions—personal, industrial, and governmental—and provides a basis for learning how excellent organizations make good decisions and applying these findings to diagnose and improve the decision-making processes of other organizations. The six elements are:

1. *Appropriate frame.* Does the decision have a clear purpose, well-defined scope, and proper perspective?

2. *Creative alternatives.* Has a wide range of significantly different yet potentially attractive alternative solutions for this decision been considered?

3. *Reliable information.* Has all available relevant information been gathered and utilized, including consideration of the uncertainty in this information?

4. *Clear values.* Has the organization clearly articulated its objectives and values, as well as how to make tradeoffs among these objectives and values?

5. *Correct logic.* Given multiple alternatives, uncertain information, and multiple objectives, have the relationships necessary to assess the value of each alternative correctly been defined and implemented properly?

6. *Commitment to Action.* Does the organization reach decisions through a well-defined decision process that ensures the right involvement of the right people and leads to a commitment to act on the recommended decision?

This framework provides conceptual guidance on what it takes to achieve a high-quality decision and can be used during the course of a decision process to measure progress. Extensive research and application experience has demonstrated that the decision quality paradigm applies to the R&D decision-making process (Matheson and Matheson, 1998; Matheson and Menke, 1994; Menke, 1994).

In 1992, the R&D Decision Quality Association set out to determine whether R&D decision making could be benchmarked and, if so, what were the best practices that leading industrial R&D organizations use to achieve R&D decision quality. The result was the framework shown in Figure F.1 of 45 best practices for R&D decision-making (Matheson, et al. 1994). These include practices for making quality decisions, organizing for decision quality, and improving decision quality. Making quality decisions further divides into practices for establishing a sound decision basis, for developing technology strategy, for managing the R&D portfolio, and for developing and evaluating project strategies. Organizing for decision quality subdivides into practices regarding organization and process, relationships with internal and external customers, and R&D culture and values.

The validity, importance, and usage of these 45 practices were tested in the benchmarking. To provide quantitative benchmarks, nearly 300 R&D organizations answered a common questionnaire

154

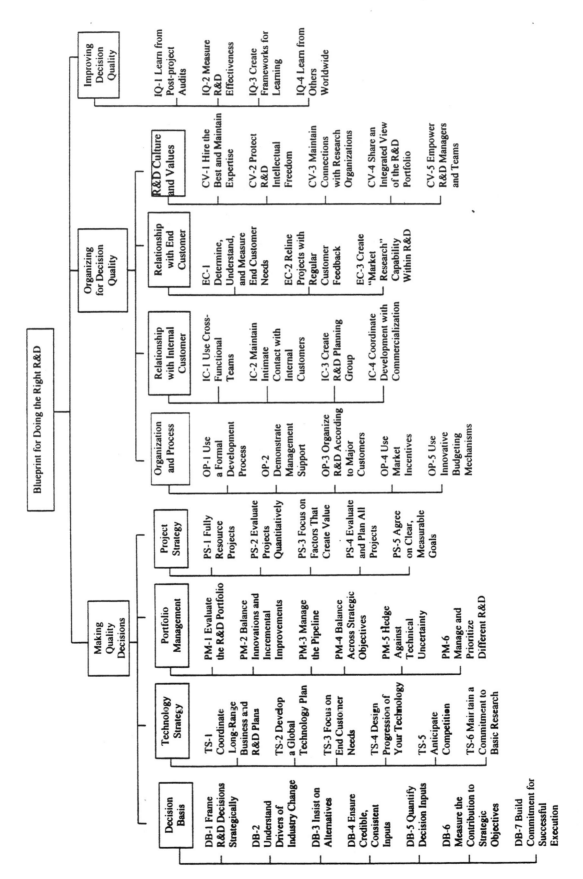

FIGURE F.1 These 45 best practices are the building blocks for an organization to routinely achieve decision quality.

regarding their usage of these best practices. The questionnaire included some overview information about the organization and the nature of its R&D, three questions about the R&D-related performance of the organization, and a matrix for documenting the applicability, frequency of use, quality of execution, and importance to decision quality of the 45 best practices. They then identified a sample of 79 highly regarded R&D organization using peer group nominations, Baldridge award winners, and *Fortune's* most admired company list, called the "best companies." The 79 organizations in this "best companies" subgroup are listed in Table F.1.

If these are indeed best practices, they should impact performance. Using the average of the three measures from the questionnaire as a performance score, the entire database of nearly 300 organizations was divided into two groups, high performers (i.e., those with a performance score above the median) and low performers (i.e., those with a performance score below the median). Figure F.2 shows that the high performers are more likely to actualize[1] best practices than the low performers. This result is the primary "objective" evidence for the validity of most of these best practices, although the fact that they were initially derived from the common behavior of outstanding organizations is also very compelling. Although the difference in actualization on any one practice is not huge, most of the best companies interviewed use a process involving many of the 45 practices in a coordinated way, thus maximizing their combined impact to achieve substantial performance advantages.

Note that the top ten practices in terms of greater actualization by high performers include four decision basis, two technology strategy, two project strategy, and two internal customer practices. The top twenty practices, whose usage discriminates high and low performers, include four decision basis, four technology strategy, three portfolio management, five project strategy, one organization and process, two internal customers, and one culture and values practice. Overall, 16 of the 20 practices whose actualization discriminates high and low performers are for making quality decisions, 4 of the 20 concern organizing for decision quality, and none of the 20 concerns improving decision quality through measurement and learning. This does not mean that the latter two areas are less important. It simply means that there is much more variability in the actualization of practices for making quality decisions than in the other two areas. This suggests that many R&D organizations can improve R&D performance and gain competitive advantage by improving the frequency of use and quality of execution of practices for making quality decisions.

Figure F.3 shows the importance of the practices in terms of their perceived potential contribution to decision quality. Potential contribution to decision quality was scored on a scale of 1 to 7. Since people often ignore the extreme high and low scores on a rating scale, practices whose average score is above 6 from a large number of respondents must be considered extremely important. Ten of these stand out for having a mean potential contribution to decision quality greater than 6.0, specifically:

1. Understand the drivers of industry change
2. Coordinate long-range business and R&D plans
3. Focus on end customer needs
4. Agree on clear measurable project goals
5. Use a formal development process
6. Use cross-functional teams
7. Coordinate development with commercialization
8. Determine, understand, and measure end customer needs
9. Refine projects with regular customer feedback
10. Hire the best and maintain expertise

These 10 practices are essential for R&D strategic excellence (Menke, 1997a).

---

[1]As used here, the term "actualization" is the product of frequency of use and quality of execution, so it measures *effective* use of the practices. Since actualization is the product of two numbers between 0 and 1, it may be more intuitive to think about the square root of actualization as the level of effective usage.

TABLE F.1   Organizations That Participated in the Quantitative Benchmarking of Best Practice Usage.

Abbott Diagnostics	Conoco
Amoco EPTG	Corning
Amoco Chemical	Daimler Benz Central Research
Amoco Technologies	Dow Chemical
Amoco Corporate Research	DuPont Central R&D
Apple Mac Systems Division	DuPont Business Unit
Asahi Glass Central Research Center	DuPont Merck Project Management
AT&T Bell Laboratories x 2	E. Merck
AT&T Operating Systems Business Unit	Eastman Chemical Research
AT&T Global Public Network Platforms	Eastman Chemical Development
AT&T Network Systems	Eaton
AT&T GBCS	GE Corporate Research and Development
Bayer Pharmaceutical	GE Aircraft Engines
Boeing Commercial Aircraft x 2	Genetech Discovery Research
Bristol-Meyers-Squibb Project Management	Genetech Product Development
Coca-Cola	Gillette Corporate R&D
Compaq Corporate Development	Gillette Research Institute
Compaq PC Product Development	Glaxo Wellcome
HP Corporate Engineering	Motorola Energy Products
Hoffman-LaRoche	Pfizer Central Research and Development
IBM Applications Development	P&G Laundry and Cleaning
J&J Medical	Rhone-Poulenc Rorer Project Management
J&J Pharmaceutical Research Institute	Sandoz Corporate Project Management
Kirin Brewery	Shell Chemical
Lilly	Shell Development
Mallinkrodt Medical	Shell International Petroleum Research
Matsushita Electrical Industrial Central Research Labs	SmithKline Beecham
Matsushita Electrical Industrial Tokyo Research Institute	Sony Research Center
Merck Project Management	Sumitomo Chemical Tskuba Research Center
Merck Regulatory Affairs	Sumitomo Electric
Microsoft Development	3M Corporate Research
Microsoft Research	3M Coporate Research Process Technology Lab
Mitsubishi Chemical	3M I, I & E Sector
Mobil Chemical	TI Central Research Laboratories x 2
Motorola Automotive	TI Systems and Software R&D
Motorola Semiconductor Products	Xerox Corporate Research and Technology x 2
Motorola T&Q	Xerox High Volume Reprographics

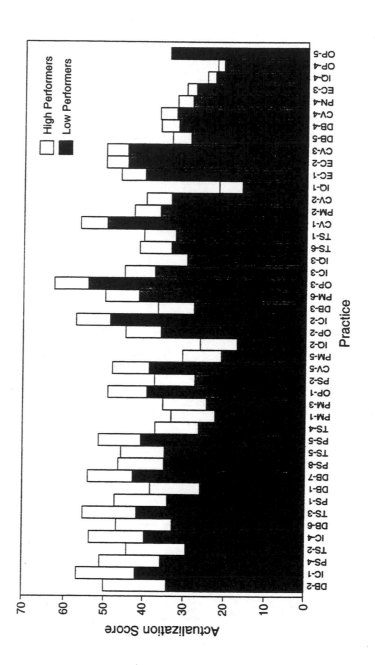

FIGURE F.2 Actualization: high performers verses low performers.

158

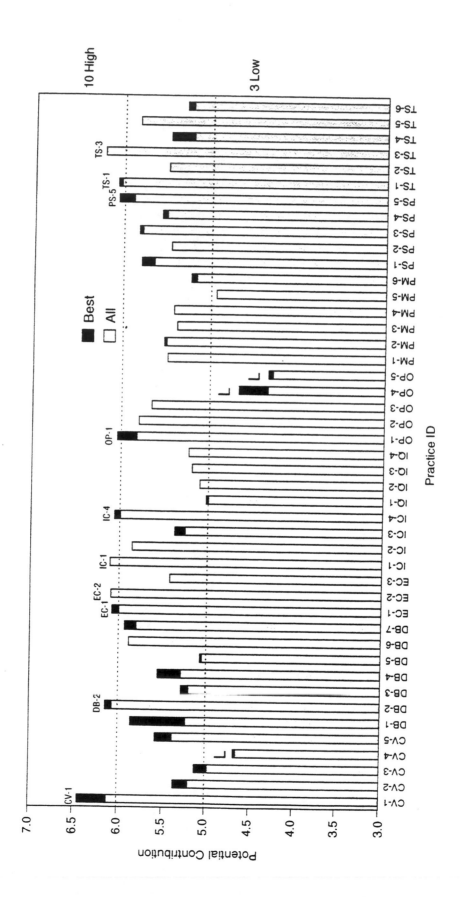

FIGURE F.3 Mean potential contribution—"best companies" verses all.

To gain additional insight, the top 10 percent of the best company data for potential contribution and actualization were looked at on a practice-by-practice basis. The mean score of this group (which is different for every practice) is called the benchmark. These benchmarks establish standards of importance and doability for each of the 45 R&D decision quality practices. Since they are the top 10 percent of a highly selected group of excellent R&D organizations, they truly set a high standard. Table F.2 lists the 79 best company mean importance and actualization of the 10 essential practices and compares them to the benchmarks.

Figure F.4 shows the mean potential contribution of the best companies versus the benchmarks (i.e., average of the top 10 percent) for each practice. The best company data are the same as presented in Figure F.3, but are ranked here in descending order. For most of these practices the benchmark is well above the mean of the best companies; for 5 of the practices the benchmark is 7.0, indicating that at least 10 percent of the best companies rated the potential contribution of the practice to decision quality as high as possible. For 30 of the 45 practices the benchmark potential contribution is greater than 6.0, which again indicates the importance of the majority of these practices.

However, despite their acknowledged importance and potential to improve performance, the actualization of these practices varies widely. Moreover, there is a very large gap in mean actualization between the best-company and the benchmarks. As Figure F.5 demonstrates, the benchmark actualization for the vast majority of practices is more than 80 percent, whereas the mean actualization for the entire best companies sample starts at 60 percent and drops rapidly to about 20 percent. For most of the practices, benchmark actualization is more than twice the mean actualization for the best companies. From the benchmarks we can conclude that it is possible to implement nearly all of these practices to a high degree. The mean actualization of the entire best-company sample shows that even many of the "best" companies have substantial room for improvement. One reason is that many of these practices are difficult to learn and implement effectively. Another reason is that many R&D organizations have a tradition of being independent from their internal and external customers, which is not very appropriate for today's competitive environment. The Strategic Decisions Group and the Quality Director's Network

TABLE F.2 Behavior of 79 Organizations Benchmarked by Ten Practices Essential for Excellent R&D Decision Quality as Indicated

Decision Quality Best Practice	Potential Contribution to Decision Quality (0-7)		Actualization (0-100%)	
	Mean	*Benchmark*	*Mean*	*Benchmark*
Hire best and maintain expertise	6.4	7.0	60	94
Focus on end customer needs	6.2	6.7	48	83
Determine, measure, and understand customer needs	6.1	7.0	46	88
Understand drivers of industry change	6.1	7.0	41	79
Use cross-functional teams	6.1	6.7	55	98
Use a formal development process	6.0	6.8	60	96
Coordinate development with commercialization	6.0	6.8	49	92
Agree on clear, measurable goals	6.0	6.7	50	91
Coordinate long-range business and R&D plans	6.0	6.3	39	82
Refine projects with regular customer feedback	5.9	6.8	45	90

NOTE: Actualization = frequency of use × quality of execution (i.e., effective usage)

160

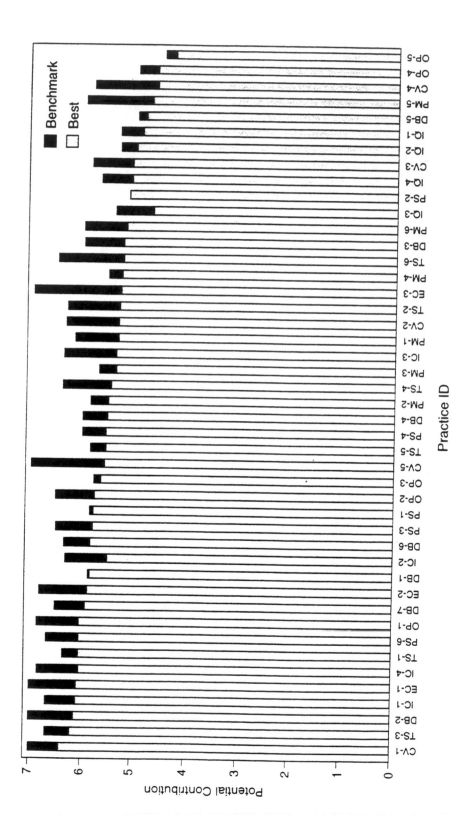

FIGURE F.4 Mean potential contribution—"best companies" verses benchmarks.

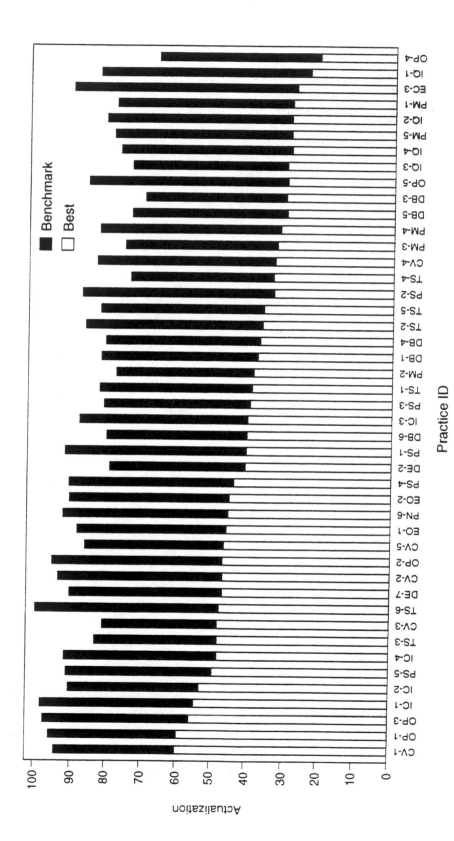

FIGURE F.5 Mean actualization—"best companies" verses benchmarks.

of the Industrial Research Institute studied this issue in the context of several member companies and identified a series of organizational enablers that assist in implementing best practices (Menke, 1997b). Further analysis has identified 10 other practices for gaining competitive advantage. These are the 10 least well actualized from a set of 28 practices judged to be very important in terms of performance impact or potential contribution (starting from the lowest actualization):

1. Learn from post-project audits
2. Evaluate the R&D portfolio
3. Measure R&D effectiveness
4. Hedge against technical uncertainty
5. Create frameworks for learning
6. Insist on alternatives
7. Manage the pipeline
8. Design progression of your technology
9. Evaluate projects quantitatively
10. Anticipate competition

Organizations that want to gain a competitive advantage can very likely do so by actualizing these practices better than their competitors. Specific data on the importance and actualization of these 10 practices are shown in Table F.3 for both the mean of the best company group and the benchmarks.

These practices derived from industrial experience were reviewed by our committee for their perceived relevance to DOE OST decision making. The committee's conclusion is that many of the same practices should be relevant to making good RD&D decisions in the DOE-EM environment.

Below is a list of the 45 best practices from Matheson, Matheson, and Menke (1994) gleaned from a survey of industrial companies with large internal R&D operations.

TABLE F.3   Ten Practices that Offer the Possibility of Gaining Competitive Advantage

Decision Quality Best Practice	Potential Contribution to Decision Quality (0-7)		Actualization (0-100%)	
	Mean	*Benchmark*	*Mean*	*Benchmark*
Lean from post-project audits	4.9	5.2	24	83
Evaluate the portfolio	5.3	6.2	29	78
Measure effectiveness of R&D	5.0	5.2	29	81
Hedge against technical uncertainty	4.8	6.0	29	79
Create frameworks for learning	5.2	5.3	30	74
Insist on alternatives	5.3	6.0	30	70
Manage the pipeline	5.4	5.7	33	76
Design progression of your technology	5.4	6.4	34	74
Evaluate projects	5.1	4.7	34	87
Anticipate competition	5.6	5.8	36	82

NOTE: Actualization = frequency of use × quality of execution (i.e. effective usage)

THE 45 BEST PRACTICES

Practice 1. Frame R&D Decisions Strategically

Distinguish clearly between strategic and operational situations, and frame each strategic decision at the appropriate level: technology strategy, portfolio management, or project strategy.

Practice 2. Understanding Drivers of Industry Change

Products that result from current R&D projects will be launched in a future business environment. It is critical for R&D planning to anticipate changes that might occur in the industry during this time. R&D management should understand all potential drivers of change (e.g., competition, technology, regulatory) and incorporate their impact in decisions.

Practice 3. Insist on Alternatives

Although it may be tempting to go along with the first alternative suggested, take the time to develop alternative ways of achieving similar objectives. Frequently this approach stimulates creativity and identifies an alternative that is clearly superior to the initial strategy.

Practice 4. Ensure Credible, Consistent Inputs

Good decision making depends on inputs that can be verified and trusted. Furthermore, for decisions across projects, inputs and assumptions must be consistent. R&D requires a methodology that allows decision makers to check all inputs and verify their credibility and consistency.

Practice 5. Quantify Decision Inputs

Although expressive, English has ambiguities that can lead to confusion, misunderstanding, and poor decisions. Whenever possible, and especially when making major decisions, quantify decision inputs to ensure clear and precise communication. Where information is uncertain or subjective, introduce ranges of inputs as well as probabilities. These practices lead to clarity, simplify analysis, and improve decision making.

Practice 6. Measure the Contribution to Strategic Objectives

It is important to measure each project's contribution to strategic objectives to ensure that it supports the business vision. By considering a clear statement of objectives, goals, strategies, and measures, you can more effectively evaluate R&D projects as well as the R&D portfolio, selecting projects that contribute the most to strategic objectives.

Practice 7. Build Commitment for Successful Execution

A good decision is worthless unless the willingness to implement it enthusiastically exists among all those whose efforts are required for success. For this reason the decision process must be fair and open, with ample opportunity for participation of all relevant individuals and functions. An important objective of the decision process is to build organizational commitment step by step, so that the decision leads naturally to effective action.

Practice 8. Coordinate Long-Range Business and R&D Plans

Coordinating R&D plans with long-range business goals and objectives can result in significant advantages. The company's strategic business vision usually incorporates information and insights from many internal and external sources--including the corporate mission statement, integrated business unit strategies, internal business planning documents, and business information obtained from major customers. With a clear, unbiased view of long-range business strategy, R&D can align its programs and projects to achieve business goals and objectives and add value to the company. R&D can also influence and improve business strategy.

Practice 9. Develop a Global Technology Plan

The greatest economies of scale are available to corporations that organize and conduct their technology development on a global basis. Great leverage can be gained by thinking and planning globally while acting locally. At every level, from comprehensive technology planning to organizational efforts involving the chief executive officer, a corporation has to act on its global vision in applying fundamental technologies to serve the needs of diverse markets.

Practice 10. Focus on End Customer Needs

Developing a technology strategy—a process that includes selecting a comprehensive set of technical alternatives for consideration and evaluation—can be made easier by focusing on end customer needs. This needs-focused approach helps ensure that the strategic alternatives are relevant to customer requirements. More important, this approach establishes clear criteria for determining the best strategy to pursue.

Practice 11. Design Progression of Your Technology

Effective technology strategy allows you to determine and manage the rate at which you will advance technology. For example, one strategy would be to use technology advances to refresh well-known project brands—a strategy that deliberately renders current product versions obsolete. Another would be to gain economies of scale by making capital-intensive investments in long-term projects that regularly produce spinoffs while building toward blockbuster innovations every few years. These practices are an example of how to achieve significant gains by strategically managing technology advances.

Practice 12. Anticipate Competition

Build models, obtain competitive analysis information, conduct surveys, and apply other techniques to anticipate the competition.

Practice 13. Maintain a Commitment to Basic Research

Given the importance of basic research to long-term success, resist attempts to cut back on basic research to create the appearance of short-term gains.

Practice 14. Evaluate the R&D Portfolio

Portfolio analysis is a key practice that categorizes, compares, and evaluates the portfolio as a whole. Compare alternative portfolios first on probability distributions on net present value. Other comparative measures should include expected product launches over time, patterns of resource requirements over time, and the ability of the portfolio to meet corporate objectives. This approach provides a clear sense of the value of the entire R&D portfolio, leading to significantly better portfolio investment decisions.

Practice 15. Balance Innovations and Incremental Improvements

A balanced portfolio gives your organization the best chance to achieve sustained R&D and commercial success. Strive for the best balance between projects that provides incremental improvements to current products and projects that have significant breakthrough potential.

Practice 16. Manage the Pipeline

Manage the product pipeline to ensure a smooth, reliable flow of products and product improvements. A hallmark of the portfolio approach to R&D management is the ability to balance long-term and short-term interests and support the technology strategy that brings the greatest overall value to the company over time.

Practice 17. Balance Across Strategic Objectives

Effective portfolio management entails aligning individual projects with the achievement of strategic objectives. This requires well-defined strategic objectives and consistent priorities that permit the appropriate positioning of projects—by category, by business unit, by technical focus. By segmenting the portfolio and managing projects accordingly, you can balance R&D across strategic priorities.

Practice 18. Hedge Against Technical Uncertainty

In managing a portfolio, protecting the company against unforeseen and unpredictable technology failures is very difficult. The best practices involve hedging the portfolio against technical, regulatory, and industry uncertainties through systematic project analyses and portfolio evaluation.

Practice 19. Manage and Prioritize Different R&D Differently

A comprehensive R&D program encompasses basic research and technology development along with projects directed at process improvements, incremental enhancements to product lines, new product development, and breakthrough innovations. Portfolio managers should be sensitive to differing objectives among projects and manage them accordingly.

Practice 20. Fully Resource Projects

When budgets or staff are constrained, it is tempting to cut from all projects rather than continue to support some projects while cutting others entirely. Unless the cuts are minor, avoid this practice. Inadequate resourcing delays projects, sometimes to the point where competitors reach the market first. Since delays usually result in higher overall costs, inadequate resourcing can result in higher total project cost. When faced with constraints, R&D should fully resource the most important projects rather than cutting across the board.

Practice 21. Evaluate Projects Quantitatively

Quantitative evaluations of projects are essential for clear communications and equitable prioritization. The best companies combine economic and technical evaluations, including market potential as well as technical hurdles. Uncertainties are addressed directly using the language of probability. Most projects are ultimately compared on the basis of the expected net present value and a productivity indicator—such as expected net present value divided by expected R&D cost.

Practice 22. Focus on Factors That Create Value

New products and processes may have many technical performance features, but their value to the company usually depends on only a few. Knowing which features add value can help you select projects and allocate R&D resources better. Project efforts should focus on features that create the most value. Terminate projects that do not meet key performance factor requirements in favor of more promising efforts.

Practice 23. Evaluate and Plan *All* Projects

Good projects can withstand a rational analysis in the cold light of day; bad ones are likely to wither. When powerful champions recommend projects, there is a tendency to move ahead without evaluation or careful planning. This can result in poorly executed research and misallocation of resources. Always plan and fully evaluate projects regardless of pressure from champions.

Practice 24. Agree on Clear, Measurable Goals

Project success requires a fully empowered team that agrees on strategic objectives, major milestones, and clear, measurable goals. The best approach is to build alignment within the team and give it the power to negotiate, obtain authorizations, and implement midcourse corrections to move R&D toward achievement of the agreed on goals and objectives.

Practice 25. Use a Formal Development Process

For R&D that leads to a sustained presence in national and international markets, the best companies use a formal development process with phases, checkpoints, and milestones to frame decisions and track their implementation. The process includes features and documentation that make continuous improvement possible through iterative refinement.

Practice 26. Ensure Active and Explicit Management Support

Management support for R&D strategy and implementation should be active, explicit, and visible throughout the company. In particular, you can benefit from the involvement of the CEO and board of directors—or senior business unit executives—in developing organizational alignment of business objectives, technology strategy, portfolio strategy, and project strategy.

Practice 27. Organize R&D According to Major Customers

By organizing according to major customers, R&D remains responsive to customer needs. Researchers and managers in each organization learn what it takes to be successful in the customer's business. Alignment between customers and technology centers grows naturally from the organization of R&D.

Practice 28. Use Market Incentives

To obtain solutions that your organization cannot or does not wish to achieve internally, offer incentives to other organizations with an interest in achieving these solutions. A variety of incentives could be introduced for this purpose, from prizes and awards, to licensing and royalty agreements that take advantage of products based on successful R&D by the outside organization.

Practice 29. Use Innovative Budgeting Mechanisms

Full-burden costing, collaborative funding, and continual zero-based budgeting are several innovative approaches to R&D budgeting that can result in improved performance.

Practice 30. Use Cross-Functional Teams

Any successful product must be successful in every phase—R&D, manufacturing, marketing, sales, and customer support—so the best project teams involve representatives from these functional areas. If the R&D team has no information about what the market wants or what makes the product easy or difficult to manufacture, the product probably will not reach its market potential; it may not even reach the market. Multifunctional teams ensure that information *from* all relevant functions is conveyed *to* all relevant functions. This practice increases the probability that R&D will create a first-rate product that meets customer needs and regulatory requirements.

Practice 31. Maintain Intimate Contact with Internal Customers

Maintaining close relationships with internal customers is a key to R&D success. R&D should communicate with customers, learn from and educate them, involve them in decision making, and organize them to benefit from R&D programs and activities. This important practice area was mentioned repeatedly in benchmarking interviews.

Practice 32. Create R&D Planning Group

To make sure that "doing the right R&D" gets sufficient attention, have your R&D organization create its own planning group. The group supports management in making the right decisions and therefore in doing the right R&D—complementing the operational approach of others who are appropriately concerned with doing the R&D right. This planning group works with R&D and business managers to keep strategic questions on the agenda and coordinate the process of making strategic R&D decisions.

Practice 33. Coordinate Development with Commercialization

Many organizations must deal with the problem of R&D that is not effectively commercialized. Successful R&D organizations take three important steps to make sure that their value added is captured by the company: they (1) use a formal development process, (2) sustain projects through this process, and (3) obtain strong commitments from their marketing organizations to commercialization of project results.

Practice 34. Determine, Understand, and Measure Customer Needs

Successful companies make every effort to determine, understand, and measure the needs of their internal and external customers. Their R&D organizations determine customer needs by soliciting frequent input. Many techniques can be used to gain customer feedback and inspire customer loyalty—from focus groups and customer associations to telephone surveys. They all have a structured interaction that returns knowledge, insight, and information to the organization.

Practice 35. Refine Projects with Regular Customer Feedback

As part of the ongoing technical planning and project development process, successful companies take advantage of a very powerful resource: their customers. In regular customer meetings, they review progress on current programs, develop new ideas, and align priorities; these inputs are then used to guide decision making. The result can be a customer-driven organization that responds to consumer needs for information, and tests products and processes with relevant customers in real situations as early as possible. For more effective R&D, use customer input to refine project concepts for an actual marketplace.

Practice 36. Create "Market Research" Capability Within R&D

In many industries, an effective R&D organization needs its own capability to do conceptual product design and market research. You need the vision that arises from direct access to customers and customer information. Focus groups, surveys, analysis of competitive information, and other mechanisms produce a more direct encounter with customers than can be provided by another organization. While your marketing organization handles activities such as advertising and sales support, the R&D organization creates the vision and generates information about what customers will need in the future.

Practice 37. Hire the Best and Maintain Expertise

The most successful R&D organizations hire the best professional staff available and provide them with opportunities and substantial support for further professional development. In this area, the best practices are those that inspire knowledge-intensive endeavors from highly motivated R&D staff members who feel that they have found the right place to make a personal contribution.

Practice 38. Protect R&D Intellectual Freedom

Companies that participated in the benchmarking study strongly endorsed the need for an open environment characterized by intellectual freedom, encouragement and protection of risk-takers, debate across an unlimited horizon of issues, conflict resolution procedures, and creative negotiation with outside departments. The practices they contributed are designed to stimulate the best possible thinking leading to successful R&D.

Practice 39. Maintain Connections with Research Organizations

To maintain technical vitality, it is important to participate in research organization networks. Active involvement of researchers and developers promotes technical achievement and advances the discovery of new technologies that may be exploited to competitive and commercial advantage. Opportunities to benefit from interactions with research exist within large organizations as well as throughout the larger community.

Practice 40. Share an Integrated View of the R&D Portfolio

The R&D culture and organization benefit greatly from sharing an integrated view of the portfolio. People who are dedicated to a single program or project need to understand how their work fits into the larger scheme of things. A major benefit of understanding the entire portfolio is that people working on individual projects can see these projects in context, understand how they contribute value to the organization, and can collaborate proactively with others to bring them to successful completion.

Practice 41. Empower R&D Managers and Teams

A variety of practices were recommended to empower project teams to accomplish their objectives. Empowerment to negotiate midcourse corrections, change direction in the search for better technological solutions, and manage the project according to its own imperatives results in more effective management and decision making because the people closest to the situation are making the decisions. What is required at the project level is also required at the levels of technology and portfolio strategy, where global accountability for success must be accompanied by global authority to direct R&D activities, or management lacks the ability to change course to achieve the best results.

Practice 42. Learn from Post-Project Audits

R&D decision makers can learn from retrospective audits of all projects that reach the market-testing phase. An audit shows how good the decision-making process is and reveals how it can be improved. Post-project audits are a key to continuous improvement in building the effectiveness of R&D teams. You can learn more from failures than you can from successes. You can learn most by finding the factors that differentiate failures from successes.

Practice 43. Measure R&D Effectiveness

A key to improvement is knowing where you stand at present, so you know where to focus your next efforts. The best organizations devote substantial attention to measuring the effectiveness and productivity of R&D. These measurements take two forms: retrospective and prospective. The best retrospective measures relate to commercial success—for example, the fraction of revenues and profits from new products. The best prospective measures relate to the position of R&D in future operations of the company—for example, expected net present value and contribution to strategic goals.

Practice 44. Create Frameworks for Learning

One of the best approaches to R&D involves creating a framework for learning—individual learning and organizational learning—and making sure that the benefits of continuous learning are fully realized. You can institute quality programs that reinforce leadership in various areas of R&D, or you can allow the documentation process to provide structured organizational learning—at the levels of technology planning, portfolio management, and project strategy.

Practice 45. Learn from Others Worldwide

In rapidly changing global markets, where uncertainty and unexpected change place added demands on expectations and success or failure, R&D organizations must be open to learning as much as possible from others worldwide. The best approach involves identifying the most successful organizations, especially competitors and organizations in other countries, understanding what makes them superior R&D performers, and adapting your version of their best practices.

Appendix G

Prioritization and Decision Making in Technology Development at the Electric Power Research Institute

This appendix summarizes information learned by a committee member (Edwin Zebroski) in interviews with EPRI managers[1] at the Electric Power Research Institute in Palo Alto, California, on September 11 and October 2, 1997. Following a brief explanation of similarities between OST, EPRI, and the Gas Research Institute, the EPRI decision-making process is discussed in more detail.

ANALOGIES BETWEEN OST, EPRI, AND GRI

The EPRI and GRI support research for their respective industries, electricity and natural gas, in market sectors such as production, distribution, environmental impacts, and new commercial uses.[2] Within each institution, the same decision-making processes for project selection and evaluation were applied to the R&D activities across all sectors, making these processes ones that could be applied generally to R&D work of many kinds (i.e., not restricted to R&D activities in any one sector, such as environmental impacts).

The committee devoted special attention to EPRI and GRI because of similarities between these institutions and OST. All three organizations formulate and manage R&D that is executed by contractors, partly on a sole-source basis and partly on an RFP and open bid basis. Other similarities exist in their respective research portfolios, organizational structures, and context.

Research Portfolios

Several features of the R&D activities of EPRI and GRI are analogous to those of OST. Both EPRI and GRI develop process improvements that lessen environmental impacts and facilitate compliance with environmental regulations. Both EPRI and GRI have programs on environmental monitoring and remediation technologies. Both manage R&D in several technical areas, with projects at various stages of technical maturity (i.e., from basic research to applied engineering and demonstration).

---

[1]The managers interviewed were Robert Brockson (Environment), Gene Eckhart (Customer Service Group), Michael Evans (EPRI CSG Co., a profit-making subsidiary), Mark Samoyj (Power Electronics and Power Systems), and Steve Gehl (Strategic Planning).

[2]These might be engendered through the increased availability of an affordable energy source.

Organizational Structure

As described in further detail in this appendix and the next, both EPRI and GRI have several tiers of program structure, each managed by staff and advised by an oversight body. These institutions serve a customer base that is a collection of quasi-independent member utilities that support a centrally coordinated R&D program based on their perceived benefits (to each member utility as well as to the industry at large).

Context

Other similarities are those of context and include the broad range of interests that these institutions must serve and the changes in recent years due to the changing interests and needs of their respective clients. Like OST, both EPRI and GRI have experienced declining and more tightly defined, controlled, and scrutinized budgets. Both EPRI and GRI have experienced greater pressures to be responsive to the short-term needs of their clients.

Features of EPRI and GRI Decision-Making Processes

Both EPRI and GRI have highly structured decision processes for goal setting, project definition, and project selection. Although the decision organization structures differ, they are functionally equivalent in providing decision processes that connect the setting of general goals and missions with the formulation and selection of specific projects.

Both EPRI and GRI have made the transition from decision processes that were largely "top-down" (in defining goals, missions, and content of their R&D programs) to more "bottom-up" processes, which develop a program content that is closely coupled to the current needs of their clients. Both organizations have good track records of the yield on R&D in terms of benefit-cost ratios for individual projects that are selected and eventually applied in the field. They also have relatively good track records in terms of the cumulative benefit-cost ratios of their entire programs for their client communities.

Further remarks on EPRI's evolving decision-making process appear below. Appendix H discusses GRI's methodology and provides tables suggestive of analogies to OST within DOE-EM.

ORIGIN AND PAST PRACTICES OF EPRI, 1972-1990

EPRI was set up by the electric utility industry in 1972-1973 as a response to a congressional proposal to establish a federal organization apart from the federal Energy Research and Development Agency (now DOE) that would develop improvements in electricity supply, delivery, and use. The federal agency was to be supported by a tax on electricity bills. The counterproposal by the industry was to form a nonprofit research organization managed by the industry and supported by dues from the electric utilities and by pass-through charges levied on the regulated earnings. This proposal was accepted by the U.S. Congress.

Most of the divisions of the new EPRI organization inherited some legacy R&D programs that had been supported by the Edison Electric Institute (EEI), a trade organization. These projects had been initiated by EEI and managed by advisory committees that met several times a year. As the size and complexity of these programs grew and as the projected 10-year expenditures rose to several billion dollars, the need for ongoing professional management and subject matter technical expertise became widely recognized in the utility industry. This recognized need coincided with pressure from the U.S. Congress for an energy research institution, which resulted in the formation of EPRI.

Funding and Membership

Initially, EPRI was funded by dues that were based on a fixed formula paid by participating utilities. The formula took into account the number of customers, the overall size of power generation, and the overall revenues of each company, and covered both stockholder-owned utilities and cooperatives. This fee provided the utility with membership in the organization, access to all R&D results, and opportunities to serve on several levels of oversight and technical committees and the Board of Directors. Member utilities included power cooperatives and municipal and federal power authorities such as the Tennessee Valley Authority. In the last few years, about 20 percent of the companies that are non-utility generators have become members. At its peak, total membership represented about 90 percent of the electric-generating capacity of the United States.

Organizational Structure

EPRI was run by a Board of Directors, which established the formulas for utility dues and the total size of the EPRI budget. Under the board were technical staff in divisions and departments.

Each of these management levels had an advisory group. Parallel to the Board of Directors, a Research Advisory Committee (RAC) was established that included top-level technical people from member utilities. The members were composed of directors of research, vice presidents of engineering departments, and senior project managers of utilities. Each EPRI division and department had its own technical advisory committee, with utility representation by specialists in the respective technical areas involved.

Top-Down Definition of Organization R&D Goals

The overall mission and goals of the organization was defined in general terms by the organization documents and further defined on a continuing basis by the Board of Directors.

Allocation of Budgets to Divisions

The Board of Directors and the RAC held quarterly meetings reviewing the work of EPRI. A special annual meeting of the two was held together. The outcome of this annual meeting was the determination of the overall budget level and the division of this budget into major topic categories such as Generation (spanning both fossil and nuclear), Transmission and Distribution, and Environment.

The divisional and departmental committees, together with EPRI staff, provided the RAC and the Board of Directors a list of proposed new and ongoing projects that defined the proposed goals and budget of each division. These projects were generally approved as proposed. Exceptions occurred when the board and RAC differed in their proposed division budgets. In such cases, the divisions facing a tentative budget reduction would have their lists of projects ranked in priority as judged by the staff. The lowest-priority projects would be dropped to meet the allocated budget.

Project Definition and Selection

Projects were defined by the technical staffs of the divisions and departments to contribute toward each division's or department's general goals and scopes. EPRI experienced growing budgets for most of its first two decades. This resulted in a significant fraction—typically 15 to 30 percent—of projects as new starts each year. Formal processes for contractor qualification and bid selection were established and used. A generic matrix of selection criteria was used, including the contractor's

- expertise,
- technical track record,
- financial track record, and in many cases, and
- ability to provide cofunding from another source.

Measurement and Control of Projects

The contracts for projects included schedules for major milestones and deliverables, as well as expenditure forecasts. The project engineer for EPRI would meet regularly with the PI for progress reviews to supplement the information received in written progress reports. Contract provisions permitted short-term cancellations or modification of contracts for reasons such as a failure to meet milestones, a lapse in budget controls, or a change in information on technical feasibility and/or expected benefits.

Changing Regulatory Environment and Deregulation in the 1990s

The current partial deregulation of the electric power industry has the effect of separating the suppliers of power generation, electric transmission, and local electricity distribution into three distinct entities with different interests, technical needs, and competitive pressures. Utility companies no longer have assured monopoly of supply and transmission in their territories. Thus, each utility is in competition with other utilities and with nonutility generating companies. The former common interest in improving the technologies and solving common problems is largely gone, since a particular improvement may be of greater value to a competitor.

As an indication of these trends, an overall assessment of dues started to show declining participation, resulting in a reduction of EPRI's budgets for R&D and staff. EPRI responded to this challenge by adopting a "bottom-up" decision process with more direct involvement of the most interested potential users in formulating R&D projects and goals.

CURRENT PROJECT SELECTION PROCESS

The practice of EPRI in the 1990s is to formulate its R&D program in a way that supports only the project work that member companies choose to fund with their membership dues.

A Reorganization That Enables Members to Select Projects

EPRI has been reorganized into "business units," which replaced the former divisional structure. Determination of the overall EPRI budget and allocation of budgets among the business units are longer primarily done at the board level.

In the current "cafeteria style" of operation, a "menu" of proposed projects is submitted to the Board of Directors with the endorsement of the senior technical advisory committee. This committee obtains inputs from the advisory committees of the business units (formerly divisions), which in turn get advice and project endorsements from the technical committees at the target level (programs or projects are now called "targets"). The board's approval of the list of targets is an endorsement only in principle, because each target approved by the board must then attain sufficient buy-in by member companies to support the proposed work.

Each participating company has the option of supporting any target (project) of interest, a group of related targets, or none. Only a few companies (about 10 percent of the client pool) now broadly support all or most of the program across the board.

The Board of Directors continues to provide overall policy guidance on goals, missions, and relative emphasis on the different areas of technology needs and opportunities. However, votes are weighted by the degree of participation of the voting company in the topic area involved. The board endorses the overall budget, which—with some exception for strategic research as described below—is composed of targets with specific funding commitments from interested utilities.

Project Definition and Formulation

Each proposed project is subjected to several evaluations.

Pre-Project Evaluations

Each proposed project (as well as some continuing projects) is formulated or reformulated by the interested business unit. In the design of a research proposal, before a project is presented to the member companies the proposed project is reviewed intensively by the Customer Systems Unit, which evaluates

- the potential market (i.e., the reasonably assured base of potential users);
- the benefit-cost expected if the project is fully successful;
- the probability of success (technical risk) that the technical proposal, if funded, would reach the stated performance goals and deliverables; and
- the risks of being able to accomplish the expected results within the proposed budget and time schedule.

The continuing discipline of requiring funding buy-in by potential users provides a continuing test of the effectiveness of these market estimates and cost-benefit calculations.

Technical Advisory Committee Evaluations

The project proposal, along with the market evaluation and the technical risk evaluation, is submitted to the advisory committee structure and to all of the potentially interested companies that may not be currently represented on the advisory structure. This process determines which companies will buy into any given project by cofunding it, in effect buying a "share" of the project. The cost of each share depends on the number of participants and may be adjusted by the relative sizes of the participating companies. It is possible for a single company to support a project unilaterally. Companies that do not participate in a given target do not have access to the results of the work.

Strategic R&D

A shortcoming of this cafeteria-style funding, which is recognized by EPRI managers, is that certain types of projects will not be pursued. These would include projects that have relatively long-term goals or projects that explore a possible opportunity that has substantial risk associated with it.

This deficiency is addressed by a category called "Strategic Research," formerly known as Exploratory Research. This is an assessment set by board policy of about 12 percent of the overall budget for "Targets of Opportunity." These have sometimes been relatively fundamental pieces of engineering, economic, or environmental research aimed at overcoming some obstacle in understanding a particular system. Other goals may be to overcome some limiting factor in a system's efficiency, reliability, or ability to provide a novel function.

Half of the Strategic Research budget (6 percent) is allocated to the individual business units. The other half is managed by a Strategic Research unit, dedicated to long-term or more speculative targets. The usual pattern is that the work of the Strategic Research unit establishes the feasibility of a novel or improved approach. If it appears to be able to reach field application, the appropriate business unit carries on further development, provided the target can find member support. This process appears to be working reasonably well to preserve some long-term advanced work, with about 25 percent of the programs lasting over a several-year period.

Other Current Practices and Features of EPRI Operations

The practices and features described below contribute to the viability of the EPRI way of supporting R&D.

Risk Analysis for Cost-Benefit Ratings

Environmental R&D has been especially market driven due to increasing regulatory requirements and air quality standards. Each proposal is subjected to a sophisticated risk analysis for the environmental factor or pollutant involved and the expected environmental benefit if the project is funded and succeeds. Research topics include air quality improvements, watershed management, and fisheries. Avoided costs are included in the cost-benefit assessments.

R&D Deliverables at Fixed Price

The companies that participate in a target are in effect given a guarantee by EPRI that the deliverable results will be provided for a promised cost and schedule. This provides a strong incentive to the companies to participate since it largely precludes the risk that their investment in supporting the project will be wasted. It also provides a strong incentive for EPRI and contractors to monitor and manage projects closely to avoid missed milestones, cost overruns, or loss of focus.

User Buy-in and Cofunded Demonstrations Ensure Applications

The pattern of carrying forward only projects that have specific user commitments has largely eliminated a concern noted in the 1980s that only about 10 percent of R&D results in some areas had widespread application in the subsequent three- to five-year period. Demonstration projects are conducted only when a major share of the funding comes from the host company.

Cost-Benefit Re-evaluation and Ranking

In addition to the pre-project market and cost-benefit evaluations, a special staff oversight function is provided to re-evaluate cost-benefit results and rankings periodically.

Information Support Services

Because of the changing paradigms of suppliers, marketing, and competition, the traditional patterns of customer support, billing, and information flows are also in need of rapid change. Specific examples are the management of real-time pricing of power supply and power quality issues such as reliability, freedom from damaging harmonics, and voltage stability. A number of utilities are also providing specialized telecommunications services. EPRI sees a market for special developments in such topics. It is expected that many such of these will be specific to a given local situation and therefore call for proprietary developments. A for-profit subsidiary has been formed to handle this area.

Appendix H

Prioritization and Decision Making in Technology Development at the Gas Research Institute

This appendix summarizes information obtained by committee members Michael Menke and Edwin Zebroski during a September 4-5, 1997, visit to the Gas Research Institute in Chicago, Illinois. The following description comes from interviews with GRI managers and from GRI publications; possible analogies to OST are sketched in Tables H.1, H.2, and H.3.

After a brief introduction to GRI, this appendix describes the major tool—the Project Appraisal Methodology (PAM)—used to assess technology development projects that are candidates for GRI funds. The PAM rankings (using a ROI measure) of technology development proposals are used by groups of senior GRI managers and advisory bodies to allocate funds. Further sections describe the management and advisory body structure, the establishment of quantitative goals for each business unit (BU), the top-down process of setting target budget figures, and the bottom-up process of proposing PAM-ranked projects to include in the budget.

OVERVIEW OF GRI MISSION AND MEMBERSHIP

GRI is a not-for-profit corporation, consisting of natural gas utility companies, pipelines, and producers, as members, engaging in RD&D activities of benefit to the domestic natural gas community (including the gas industry and consumers). The technology development work is done under regulatory approval and in nonproprietary venues. This is changing in some instances, and proprietary RD&D management is now one of GRI's offerings to its stakeholders.

GRI's mission is to "deliver high-value technology, information, and technical services to gas and related energy markets through cooperative research, development, and commercialization (RD&C)" (GRI, 1998; see also http://www.gri.org). GRI's RD&D program, as a result of an early 1998 settlement conference (http://ferc.fed.us/news1/pressreleases/gri.htm), is now divided into two parts: (1) a *core* gas consumer benefits RD&D segment designed to produce broadly dispersed benefits across all customer classes, and (2) a *noncore* mutual benefits RD&D program designed to produce benefits for gas consumers *and* the gas industry. By the year 2001, GRI's entire RD&D program will be in the core area. GRI also is pursuing funding not approved by the Federal Energy Regulatory Committee for (1) consortium and proprietary RD&D from investors (including but not limited to the gas industry) and (2) public goods RD&D funding from federal and state RD&D agency sources and public utility commission and city-approved gas customer funds.

The benefits to GRI membership for a natural gas company are leveraged RD&D dollars, participation in the GRI advisory structure, and easier access to the technology development projects that emerge from RD&D efforts. Most major natural gas companies are members of GRI, with the exception of Exxon, perhaps because Exxon's internal R&D is so large that the leveraging benefit is not significant enough to be worthwhile or because of an Exxon policy against consortium R&D funding.

TABLE H.1   Analogies in External Oversight and Advisory Bodies

GRI	OST
External Regulator:	
Federal Energy Regulatory Council (FERC)	Congress and all regulatory bodies (EPA, states, etc.)
External Advisory Groups[a] *(three levels):*	
Highest Level	
Board of Directors	EM-1's TAC (formerly Dr. Frank's Board of Directors)
(CEOs from member companies)	(Top-level DOE site managers, Assistant Secretary for EM and DASs from EM-30, 40, 50, and 60
	Unofficially, includes input from all other external auditors such as congressional staffers, EMAB, GAO, NRC committees, SLC and others
Intermediate Level	
PAGs for each BU	Programmatic reviews of each Focus Area and Crosscutting Program
Lowest Level	
TAGs for each BU	Technical evaluation by Independent Review Teams set up by each Focus Area and Crosscutting Program

[a]Members of these bodies are external to personnel staff of the technology development program office (except for the EM-50 DAS on the TAC and EM-50 personnel on review teams).

TABLE H.2   Analogies in Program and Management Structure Within the Technology Development Organization

GRI	OST
Top-level EMC (CEO and Senior Management)	Senior HQ OST managers (DAS and close advisors)
BU (General Manager and additional staff)	Focus Area, Crosscutting Program (Manager and additional staff)
Sources of information on market use: staff who make contacts with vendors	STCGs and needs assessment activities

GRI's competition (and its potential partners and strongest supporters) is the internal RD&D shops of member companies, which do shorter-term demonstration or technology support projects than are funded through GRI. Other competition comes from service companies developing proprietary tools, RD&D firms, and other consortium RD&D organizations. Of course, these organizations are also GRI's major partners, in many cases. Technology development at GRI is available to all members (the emerging proprietary program excepted).

THE GRI DECISION-MAKING TOOL: PAM

The GRI PAM tool subjects each technology development project to an assessment according to four quantitative measures:

TABLE H.3   Additional Parallels

GRI	OST
Who is the internal customer (of technology activities)?	
TAGs and PAGs	DOE work force of RD&D personnel, primarily in national labs
Who is the external customer[a] (of technology development activities)?	
Member utilities and commercial vendors	Users of products at DOE sites, especially DOE site managers and contractor employees
Who is the ultimate, far-end customer?	
Consumers of natural gas products	Congress plus all external auditors with global interest in cleanup at low cost (including the U.S. taxpayer)

[a]Since these external customers operate independently of each other, coordinated R&D makes sense.

1. cost savings to the gas industry resulting from implementation of the technology product (reported as a O&M cost reduction or gas price savings);

2. cost savings to the gas consumer resulting from use of the technology or product (a consumer savings in terms of capital, O&M, or operating costs; or added productivity value);

3. energy savings (British Thermal Units [BTUs]); that is, the amount of energy resource (domestic or imported natural gas or an alternative energy source) conserved due to the use of that more efficient technology or product (these calculations compare the new proposed product to the baseline [present-day] technology for a projected year of use); and

4. new and retained load; that is, for GRI's mutual benefits *(noncore)* program, consumer options enhanced,[1] or increased gas use or retained gas use in comparison to other fuels. This criterion is not used for the *core* (gas consumer benefits) program.

A fifth and qualitative measure is provided by a score sheet that addresses such issues as environmental impacts, regulatory issues, and safety-related and operational improvements that cannot be readily quantified in dollar or Btu terms.

Calculating Projected Benefits

Each proposed project is subjected to calculations that assign a dollar or Btu value to its benefits in each of these four major areas (and quantitative values in the fifth area). Business units collect market data to gauge the market (to form the estimate of market penetration for a proposed new technology). This approach uses a reference database that is maintained and used providing a "level playing field" for fuel prices, market potential, and economic parameters (e.g., discount rates) by the BUs that compete for funds within GRI (GRI, 1996b). The benefit calculation accounts for technical and market uncertainties with multiplicative factors, and uses a simple discount rate to deduce a net present value of future benefits. The five calculated benefits (estimated against the aforementioned five PAM measures) are combined statistically (GRI, 1996b) with weighing factors to produce a single quantitative value that represents the total weighted benefit of the technology development project.

---

[1]Since industrial efficiency is, in principle, not of concern to the regulator, this measure is translated into one of relevance to the consumer, by passing along such efficiencies as a lower surcharge for gas services, a measure that in the market would make gas energy displace other energy sources.

RD&D cost calculations are also done for each technology development project. These consider only the costs borne by GRI; hence, leveraging dollars with other organizations is accounted for quantitatively, by reducing GRI costs (GRI, 1996b).

With both benefits and costs calculated, a benefit-cost calculation, representing the ROI for each project (GRI, 1996b) per dollar of RD&D investment, is made for each technology development project. Proposals are then ranked according to their benefit-cost ratio. The cutoff between funded and nonfunded projects is determined essentially by the budget for technology development, with some allowance for advisory bodies to select based on the worthiness of projects that are near the funding cutoff line.

Four categories of monetary returns (benefits or costs) are noted by GRI managers. Returns can be achieved as:

1. enhanced revenues through increases in gas use (noncore only),
2. a cut in capital costs (one-time savings),
3. a cut in O&M costs, or
4. reduced costs to the consumer

for each proposed RD&D project.

The PAM estimating methodology (and the database of information kept) makes record keeping easy for retrospective analyses (e.g., to explore in hindsight why some projects were thought worthy or unworthy, or simply to explore the basis for a particular decision that was made at some point in the past). Further details on this methodology can be found in GRI literature (see GRI, 1996b, 1997b).

GRI ORGANIZATION AND MANAGEMENT

External Regulation: FERC

In a regulatory process established in the mid-1970s, the overall budget of GRI is established in conjunction with the external regulator, FERC. By regulatory mandate, a one-year program and five-year R&D plans are filed and approved. An approved plan provides a mechanism to collect an approved amount of RD&D funds for the filed year from surcharges to customers of natural gas pipelines (a "pass-through" mechanism), enabling these funds to be collected with essentially no risk to the member companies.

Advisory Structure and the GRI Employee Counterparts

A body of top-level managers and executives of member utilities constitutes GRI's Board of Directors, which has the opportunity to solicit input from several subordinate advisory bodies (see Figure H.1). The Board of Directors advises the senior GRI management group called the Executive Management Committee (EMC), consisting of the president, two senior vice presidents to whom the five BUs report, and a few other senior executive officers.

The advisory bodies immediately below the Board of Directors are three Program Executive Committees (PECs), which set budgetary guidance, and three others: the Advisory Council, the Research Coordination Council, and the Municipal Gas System Advisory Committee. The PECs represent each of the major industry sectors: LDCs, pipelines, and producers.

A separate, more tactical-level Project Advisor Group (PAG) advises each of the six GRI BUs: Supply, Transmission, Distribution, End Use Efficiency, Market Analysis and Information, and

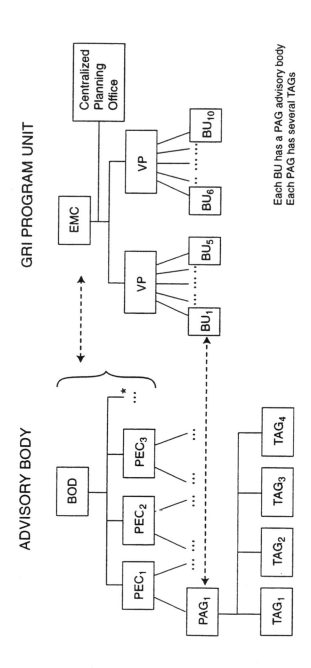

FIGURE H.1 Mapping between the GRI program units and the GRI advisory body structure. The program units interact with advisory bodies at various levels in the formulation of proposals for new R&D initiatives.

Environment and Safety. Within each PAG, several Technical Advisor Groups (TAGs) are formed around technical topics (e.g., the PAG for the Transmission BU has TAGs for corrosion issues, nondestructive evaluation, measurement, and compressors). The PAG thus has the job of ranking projects and advising the allocation of funds among projects from various TAGs, according to guidance from the PEC. The business units manage a portfolio of RD&D projects, grouped into project areas.

R&D OBJECTIVES

In this multitiered organization, different objectives are established for different levels. At the corporate level, objectives are qualitative statements that define the direction of the RD&D efforts (GRI, 1997a). Objectives are written at three levels: overall, program, and project (GRI, 1997b). Business units and senior management work together to derive quantitative goals from the objectives and use the goals to track and measure progress (GRI, 1997b).

Meeting these goals then becomes a main part of each BU's strategy. These goals are of the form of targeted improvements over the next 5 to 10 years, for example, to achieve a certain cost savings or a certain load reduction (in gas consumption per useful Btu output) or, for the mutual benefit program, increased or retained load. These quantified goals are met by delivering products, processes, and techniques to the marketplace in a certain timeframe. These goals define the impacts of technology development projects that are considered relevant for GRI funding.

As an example, the Supply BU has an objective to increase the quantity of natural gas available from emerging resources at competitive prices. Technology development projects would be funded (and their commercialization success tracked) to meet this goal. This BU is funded at $25 to $35 million per year, and employs 10 to 15 professional staff.

ANNUAL PLANNING CYCLE

GRI's practice is in principle simple—to find the technology development projects with the highest ROI, as calculated by the PAM tool. This is done each year with a top-down process of tentatively dividing the total budget among BUs, followed by a bottom-up process of generating and defending PAM-scored projects. Each year, new project ideas compete with existing projects according to PAM calculations, to determine the suite of GRI's R&D activities.

Top-Down Funding Allocation to Business Units

The top-level Board of Directors meets at regular intervals to provide guidance that senior management (the EMC) carries out. The board determines the overall annual budget and provides broad-brush, strategic, "policy"-type impressions or directions of critical needs, threats, and opportunities for the gas industry. The EMC translates this guidance into target levels of funding for major strategic initiatives.

Each BU is charged with coming up with a suite of high-ROI technology development projects to receive these funds, in a bottom-up process described below.

All but one business unit is geared toward producing results for use in the marketplace; that is, the end products are hardware process, software, or information products whose commercial sale and use make natural gas operations cheaper, safer, and/or more efficient. The exception is the Environment and Safety unit, which is geared toward activities of risk and cost avoidance (e.g., providing information to allow industry, regulators, and the general public to make informed decisions on proposed environmental legislation, or engaging in activities to prevent potential public health and safety catastrophes, accidents, or threats from occurring). Currently, the amount of funding for this business unit is set as a percentage of the total pie (the Environment and Safety amount is set by guidance as 7 to 9 percent of the total). In the near future, the Environment and Safety unit will compete for its share of the pie in the same way other business units do, by PAM ROI calculations. To do this, the Environment and Safety group is establishing methodology to quantify measures such as risk (and accident) avoidance, avoided costs, and cleanup liability and to translate these results into dollar figures.

Bottom-Up Project Initiation Process

The TAGs and PAGs of each BU are some of the places in which new RD&D ideas are generated and championed. PAG members are essentially gas company managers who have their own technical problems to solve (to achieve their own goals), and one way to help their cause is through funding relevant GRI projects. TAG members are knowledgeable about the state-of-the-art in their technical fields and are primed to champion new developments.[2] TAG members are individuals at the project manager level or the equivalent in their company; that is, they are engineers with technical problems to solve and have technical expertise as well as the ability to network within their employer organization to obtain data needed for market estimates. PAM calculations require both technical information and market assessments. The responsibility for getting quantitative estimates falls to the GRI BUs that want to champion an idea, which they forward as a PAM-scored proposal to the PAG level.

Each PAG and BU forwards to the PEC a collection of proposed projects, showing in effect what RD&D could be done with available and additional funds. The ROI numbers are used to rank the technology development proposals within and across each of the six BUs.

Subsequent EMC and PEC reviews of these proposals make funding adjustments, in effect debating over projects that are close to the funding cutoffs between BUs. A project with a lower ROI can be funded over one with a higher ROI only if BU personnel can make a compelling case for funding such a project (presumably the project's worth was poorly captured in the quantitative PAM methodology). The Board of Directors reviews the final budget for approval.

R&D PROJECT MANAGEMENT

RFPs and Contracts

Once the EMC approves funding for a proposed project, by virtue of its ROI-based ranking, RFP (or sole-source) specifications are written (by technically competent personnel) and the PI of the best-bid proposal receives that RD&D money. All of the above-mentioned R&D planning activity happen prior to issuing an RFP.Contract incentives are used in many cases to provide the contractor with incentives to achieve performance goals and seek a commercial partner.

Tracking Projects

Funded projects are tracked using the RD&C stage-and-gate process. For PAM calculations, the probability of technical success of a project was determined simply by its stage level, with each stage level assigned a certain probability of technical success (e.g., stage 2 dictates the use of a probability of x

---

[2]Other sources of RD&D ideas are GRI project managers, the manufacturing and RD&D communities, universities, and many others.

percent; stage 4, *y* percent, as a uniform practice used by all business units, for all projects at that stage). Gate reviews are done by personnel at the lowest possible management level, so as not to make the process unduly burdensome and unnecessarily formal. A few gate reviews (such as gate 2 [research initiation] and gate 4 [product development initiation] reviews), representing critical decision points, are done by more senior management. One important lesson learned to date by GRI managers is need to get marketing and manufacturing information, not just technical performance data, into gate reviews, especially by gate 4. The RD&C stage-and-gate tracking system was described as a helpful discipline, enabling a few projects that lacked potential to be recognized and terminated in early stages.

Member Company Management

It was stated that it is easier in general to sell short-term projects to advisory bodies, perhaps because of their near-term focus. Hence, guidance is established to provide program balance between near-term, midterm, and long-term RD&D, to prevent higher-risk, higher-payoff projects from being detrimentally impacted by the near-term focus of some.

How to Manage a Business Unit

According to one GRI general manager, managing a BU involves the following efforts:

1. participation in industry-wide meetings, to discuss and gain perspective of gas industry and consumer issues at several levels;
2. periodically convening GRI workshops or conferences designed to facilitate long-range thinking in the industry and generate associated new RD&D ideas, based on explorations of future major issues and potential technical solutions to them; and
3. visiting member companies and reviewing their use of GRI technologies to identify what GRI products are valued and inform the companies of where GRI is or might be headed in the future.

Each of these actions uses the interactive session as a tool to encourage the generation of new ideas within the community served by the BU.

Market Use and Commercialization

Performance is measured by BU personnel polling vendors in the commercial sector (who are obligated by contract to provide GRI with statistical information) on how many of the new devices are sold and/or used. These data are used in GRI calculations to deduce whether a goal has been met. The performance of each business unit is thus tied to the suite or portfolio of projects, rather to any particular one. Hence, individual technology development projects that are poor performers can be tolerated, especially if they are noted soon enough (at lower gate numbers), prior to heavy funding levels, and a few deliverables to the marketplace that are poor performers can be tolerated as long as enough are successfully commercialized. The measurable against which GRI business units score themselves is not how many specific technology development products are delivered to their customer or member base, but rather the total, *ultimate market impact* on gas consumers and the gas industry of new developments. On one hand, it can be argued that the technology development shop has little control over market consequences; nevertheless, market impact is the measure adopted that leads BU personnel to audit vendors to obtain data on how often new technology is used.

To pursue the above example, the Supply BU would develop improved surveying tools and techniques (to probe new reservoirs) and ask users how often such tools were used and how useful they were. The aim is to construct from such information an assessment of value (in dollars) of attaining an extra several feet of survey from the well bore shaft, which can be recast as an *avoided* cost of drilling unnecessary wells, plus the value of the additional gas reserves brought into play.

Environment and Safety Unit

This BU has a substantial risk management program to calculate a threat-to-cost ratio, analogous to a benefit-to-cost ratio. Benefits are quantified as future avoided costs, which in practice are harder to sell to advisory groups than nearer-term costs savings or new or retained load.

The ranking of project proposals within the Environment and Safety BU is done judgmentally by GRI management staff, with the help of the unit's PAG. The benefit of giving control to advise is that they take ownership of the results as well.

Proposals are packaged to appeal to all (or as many as possible) gas industry sectors, each of which would like information on how the proposed work is to be used and how it would save resources (e.g., by reducing insurance costs or reducing manpower requirements). The PAG votes, with votes normalized by business sector (meaning, presumably, that if twice as many Pipeline people come to the meeting as Supply people, the Pipeline people's votes count half as much).

One challenge of this Environment and Safety unit is that it crosscuts as the technical areas of the business units, creating a situation in which Environment and Safety unit personnel in effect report to all PAG PECs (i.e., they have many masters), which complicates management.

The Environment and Safety unit's proposals are expected to be competitive with those of the applied RD&D BUs for two reasons. The first is that an RD&D program with perceived benefits to many BUs garners many more champions. The second reason, as a manager of one of the other BUs pointed out, is that there is a generally strong awareness that this work must be done to prevent future potential liabilities to the gas industry, which are borne ultimately by the consumer in terms of higher prices for the commodity or service. It is reasonable to think that safety-related improvements would naturally be a basis for collaborative RD&D efforts (since all member companies, the consumers, and the industry as a whole benefit by safer operations).

What RD&D Does Not Get Done via PAM?

There are some intuitively important RD&D ideas, for which benefits are hard to measure, quantify, and defend, that, hence, may be unfunded by an organization such as GRI (or any similar organization that performs coordinated R&D for a large stakeholder and customer base) that uses PAM-type benefit-cost calculations to govern a corporate way of making RD&D decisions. Should such projects be done by governments, using public funds to complement what the corporate world does not do? The issue for GRI and for its stakeholder and customer base is whether any part of GRI's mission of doing RD&D for the benefit of the natural gas industry and consumers is left unfulfilled under PAM-type selection criteria for new projects. PAM also does not perform portfolio analysis, balancing such things as

- near-term versus long-term R&D, and
- supply versus gas use versus operations R&D.

Appendix I

Prioritization and Decision Making in Environmental Technology Development at DuPont

In the early 1990s, the E.I. DuPont De Nemours and Company (DuPont) adopted several goals to improve the environmental performance of its facilities. One such goal was a 60 percent reduction in air emissions of toxic substances (i.e., to reduce the poundage of toxic effluents released to the air to 60 percent of the early 1990's baseline level) from its domestic facilities by 1993 (DuPont, 1993). To achieve these environmental goals, a coordinated effort at technology development was initiated. This process is overviewed briefly here to highlight key features and characteristics and to illustrate the roles and decisions of various personnel.

This appendix summarizes information learned by committee member Michael Menke during an October 30, 1997, visit to the DuPont corporate headquarters in Wilmington, Delaware, where the Manager of Environmental Effectiveness, H. Dale Martin, was interviewed. The following description comes from this interview and from relevant publications (DuPont, 1993; Martin, 1994, 1997).

DUPONT CORPORATE STRUCTURE

DuPont's corporate structure is similar to that of GRI, in that although centralized planning and management offices exist, the individual business units are, to a large degree, autonomous in their responsibilities to the company for profit and loss, overall performance, growth, and issues of safety, health, and environment (SHE). The business units have the budget authority for funding any environmental R&D work they choose to perform.

DUPONT CORPORATE ENVIRONMENTAL PLAN

Major steps in DuPont's annual planning cycle are described here.

Top-Level Goals Imposed on Business Units

A corporate environmental improvement plan was drafted by top-level executives, due to concerns about potential regulatory penalties, a potential negative public image of the company, and the desire to be competitive in a global economy, all of which dictate sound environmental stewardship. This plan stated corporate environmental goals in general terms, such as the reduction of air toxics by 60 percent. Details of how the goals would be met ere not specified, but the goals were stated in well-defined and measurable terms, such that programs of work could be crafted to meet them. The following steps of the environmental technology development activities are described using this goal as an example.

Bottom-up Initiatives Proposed to Meet Goals

The individual business units within DuPont were then tasked with evaluating the performance of their plants, facilities, and operations with respect to environmental releases (to land, air, and water), to identify opportunities for reductions. For instance, all air emissions at transfer points and processing facilities were measured to establish a benchmark level, and methods (called "initiatives" within DuPont) to reduce these emissions were identified and costed within each business unit.

Role of Central Environmental Planning Office

The data for all initiatives were gathered by the central environmental planning office. Initiatives were sorted into broad areas such as wastewater treatment, ground water protection, thermal oxidation, and chemical engineering, and further subdivided into subcategories within these four areas. Central planning personnel were then able to identify those proposed technology development projects that could provide the maximum environmental benefit at the least cost (i.e., those proposed technology development projects that would impact the greatest number of initiatives). For example, this environmental improvement exercise identified the greatest number of business unit initiatives (146) in the subcategory of "spill containment diking," with the implication that a coordinated technology development project in this area could have multiple applications across the DuPont complex (Martin, 1994).

The next step was to identify and remove from further consideration any initiatives that did not contribute to corporate environmental goals. These were left to the individual business units to fund at their discretion and for their own purposes, presumably to help meet other goals and expectations of the business units.

The remaining initiatives were prioritized according to two measures: their implementation cost and the implied cost per unit pollutant (e.g., the initiative's cost divided by the pounds of air toxics that were prevented from being released into the environment). The results of this exercise can be displayed graphically (see Figure I.1) (Martin, 1997).

A sufficient number of these initiatives (specifically, the ones that scored most favorably against these two measures, representing the lowest-cost approach to achieving the desired goal) were funded to meet the overall goal of a 60 percent reduction in air toxins. Not all of the funded initiatives were large-scale R&D projects; some were simply procedural or process modifications, or minor modifications that the plant's local "area technical operator" could engineer and implement. Managers were held accountable for results—that is, for successful implementation of the technologies and techniques represented in the suite of funded initiatives.

A DuPont "Best Practices" manual was developed as a resource to the business units, to catalog how DuPont facilities would construct and implement environmental technologies (e.g., a common design for such technologies as scrubbers, air strippers, and soil vapor extraction techniques). This manual establishes a technology baseline to define DuPont methods for preventing, minimizing, or remediating environmental releases. Commonality of practice also helped ensure that a technology development project resulting in an improvement to a process or practice would be applied straightforwardly to multiple, identical systems. This manual was the major method for achieving leverage across business units, in that the implementation of the appropriate best practice(s) would be the preferred way for a business unit to meet its environmental stewardship objectives, rather than have the business unit inaugurate any new technology development project. Initiatives that were selected and funded to completion would be added to the Best Practices manual.

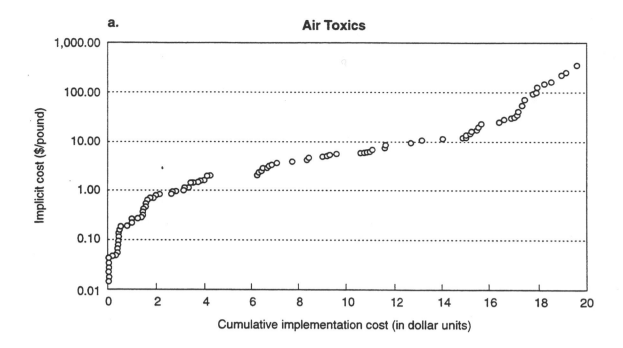

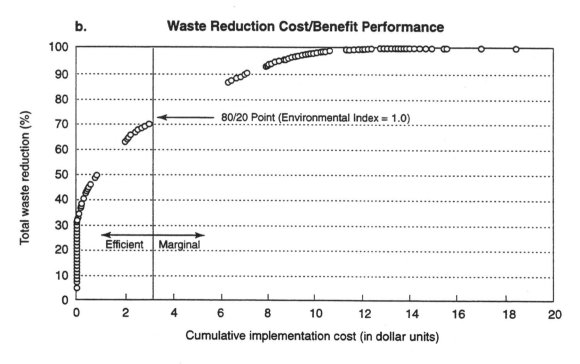

FIGURE I.1 (a) Prioritization of initiatives to reduce air toxins. Each initiative is represented by a circle and ranked by the cost per pound of effluent saved (vertical axis). If initiatives are implemented in the order in which they are ranked, the horizontal axis shows the summed cost to DuPont to implement all initiatives up to the current one, as a cumulative implementation cost, on a scale proportional to dollars. (b) Reduction in amount of waste generated as a function of cumulative implementation cost for all initiatives designed to reduce air emissions of toxic substances. This curve enables DuPont planners to select the suite of initiatives that represent the least-cost approach to achieve the greatest reductions in air emissions and to identify which initiatives are of marginal utility in meeting this corporate environmental goal. SOURCE: Martin, 1997.

Significant Characteristics of the DuPont Process

Two major attributes of the DuPont process are the following:

1. *Integration with business plans.* The environmental planning work was integrated with the business plans. Business units supplied the SHE office with their projected operations, detailing the facilities and the amount of chemical products to be generated in a future timeline. These plans established a baseline of operations from which opportunities for environmental improvements could be derived.

Further integration with business practices comes from company practice of making worker safety and health considerations integral to the way DuPont does business. Because environmental considerations are often related to these of safety and health, SHE people working to make those connections explicit enabled the environmental improvement program to receive visibility within the company.

2. *Cost-benefit analysis of environmental initiatives.* The prioritization of environmental initiatives according to cost and cost per unit effluent saved was a convenient way for DuPont business units to achieve top-level goals of upper-level corporate management, such as the 60 percent reduction of air toxics. The businesslike approach was convenient to the efficient implementation of the requisite activity to meet these goals (at minimal cost).

Implementation of the DuPont corporate environmental plan created a heightened awareness of environmental issues in business planning. Among other benefits, environmental releases were cast in business terms. That is, a pound of material released represents a pound of raw material or intermediate or final product that is wasted, making environmental cleanliness commensurate with improved cost-benefit efficiency of plant operations. The result of such awareness was to convert environmental remediation and prevention measures into business terminology and to treat them as efficiency improvements.

Other Findings

A Convenient Quantitative Figure of Merit Useful for Planning Purposes

In remediation of existing environmental releases, a quick figure of merit was derived to help identify the worst of the problems. This information helped in the decisions of which environmental R&D proposals to fund. The figure of merit was a "hazard score," the sum of three scores given based on the

- toxicity of the contaminant,
- distance to ground water (perhaps this would be better cast as a travel time, since then it would account for the permeability of the subsurface media), and
- distance to the border of the company property (again, this might be better cast as a travel time for the same reason).

Of course, a rigorous site risk assessment calculation would consider these factors in an integrated manner, but in the absence of such a calculation, a quick numerical rating of this kind was found to be a useful planning aid.

Technology Development as a Business Venture

A separate DuPont business unit was engaged in environmental remediation technology as a business opportunity, but this business unit was sold off in the mid to late 1990s due to the lack of a global market for such services. As a result, DuPont's environmental technology development is directed toward internal improvements only.

Appendix J

List of Acronyms and Abbreviations

ACNFS	Advisory Committee on Nuclear Facility Safety
ADS	Activity Data Sheet
AEA	Atomic Energy Authority
AL	Albequerque
ANL	Argonne National Laboratory
ASME	American Society of Mechanical Engineers
ASTD	Accelerated Site Technology Deployment
BEMR	Baseline Environmental Management Report
BRWM	Board on Radioactive Waste Management
Btu	British Thermal Unit
BU	Business Unit
CAB	Citizen Advisory Board
CC	Crosscutting Area
CEO	Chief Executive Officer
CERCLA	Comprehensive Environmental Response, Compensation, and Liability Act (or "Superfund")
CGER	Commission on Geosciences, Environment, and Resources (NRC)
CLN	Community Leaders Network
CMST	Characterization, Monitoring, and Sensor Technology (OST Crosscutting Program)
CP-5	Chicago Pile-5
CRADA	Cooperative Research and Development Announcement
CST	Crystalline Silicotitanate
DARPA	Defense Advanced Research Projects Agency
DAS	Deputy Assistant Secretary
D&D	Decontamination and Decommissioning
DDFA	Decontamination and Decommissioning Focus Area
DEAR	Department of Energy Acquisition Regulation
DNAPL	Dense Nonaqueous Phase Liquid
DNFSB	Defense Nuclear Facility Safety Board
DOE	U.S. Department of Energy
DOE-EM	DOE Office of Environmental Management
DOE-OR	DOE Oak Ridge National Laboratory Office

DOE-SR	DOE Savannah River Operations Office
DOIT	Develop On-Site Innovative Technologies (WGA program supported by OST)
DU	Decision Unit
EEI	Edison Electric Institute
EIS	Environmental Impact Statement
EM	Office of Environmental Management (DOE)
EMAB	Environmental Management Advisory Board (DOE)
EMC	Executive Management Committee
EMSP	Environmental Management Science Program
EPA	U.S. Environmental Protection Agency
EPRI	Electric Power Research Institute
ER	Environmental Restoration (commonly known by its EM mail stop, EM-40)
ESP	Efficient Separations and Processing (OST Crosscutting Program)
FA	Focus Area
FAR	Federal Acquisition Regulation
FERC	Federal Energy Regulatory Committee
FETC ·	Federal Energy Technology Center, co-located in Morgantown, WV (METC) and Pittsburgh, PA (PETC)
FFCA	Federal Facility Compliance Act
FSU	Former Soviet Union
FY	Fiscal Year
GAO	U.S. General Accounting Office
GRI	Gas Research Institute
HLW	High-Level Waste
HQ	Headquarters (DOE headquarters in Washington, D.C.)
HTDC	Hanford Technology Deployment Center
HTI	Hanford Tank Initiative
ID	Integrated Demonstration
IMS	Integrated Master Schedule
INEEL	Idaho National Engineering and Environmental Laboratory
IP	Industry Program
IRB	Internal Review Budget
IRTC	Interstate Regulatory Technology Council
ITSR	Innovative Technology Summary Report
LANL	Los Alamos National Laboratory
LDR	Land Disposal Restriction
LDUA	Light Duty Utility Arm
LLNL	Lawrence Livermore National Laboratory
LLW	Low-Level Waste
LOM	Lead Office Manager
LSDP	Large-Scale Demonstration Project
M&I	Management and Integration
M&O	Management and Operations

METC	Morgantown Energy Technology Center
MWFA	Mixed Waste Focus Area
NAE	National Academy of Engineering
NAS	National Academy of Sciences
NEXI	Nuclear Expertise Inc.
NPS	National Priority System
NRC	National Research Council
NSF	National Science Foundation
NTW	National Technical Workgroup
OECD	Organization for Economic Cooperation and Development
OMB	Office of Management and Budget
ORNL	Oak Ridge National Laboratory
ORR	Oak Ridge Reservation
OST	Office of Science and Technology (commonly known by its EM mail stop, EM-50)
OTD	Office of Technology Development (pre-FY 1996 name for OST and EM-50)
OTI	Office of Technology Integration
PAG	Project Advisor Group
PAM	Project Appraisal Methodology
PBS	Project Baseline Summary
PEC	Program Executive Committee
PEG	Program Execution Guidance
PEM	Program Execution Manager
PETC	Pittsburgh Energy Technology Center
PHMC	Project Hanford Management Contract
PI	Principal Investigator
PLM	Product Line Manager
PNNL	Pacific Northwest National Laboratory
PRA	Probabilistic Risk Analysis
PRDA	Program Research and Development Announcement
RAC	Research Advisory Committee
RCRA	Resource Conservation and Recovery Act
R&D	Research and Development
RD&C	Research, Development, and Commercialization
RD&D	Research, Development, and Demonstration
RDS	Risk Data Sheet
RFI	Request for Information
RFP	Request for Proposals
RL	Richland
ROA	Research Opportunity Announcement
ROD	Record of Decision
ROI	Return on Investment
RSI	Institute for Regulatory Science
SA	Strategic Alliance

SARA	Superfund Amendments Reauthorization Act
SBIR	Small Business Innovation Research
SCFA	Subsurface Contaminants Focus Area
SDG	Strategic Decisions Group
SHC	Stakeholder Coordinator
SHE	Safety, Health, and Environment
SNF	Spent Nuclear Fuel
SNL	Sandia National Laboratory
SOP	Standard Operating Procedure
SRS	Savannah River Site
SSAB	Site Specific Advisory Board
SSEB	Southern States Energy Board
SSO	Source Selection Official
STCG	Site Technology Coordination Group
STP	Site Treatment Plan
TAC	Technology Acceleration Committee
TAG	Technical Advisor Group
TD	Technology Development
TDI	Technology Development Initiative
TDRD	Technology Development Requirements Document
TEC	Technical Evaluation Committee
TEM	Technical Exchange Meeting
TIM	Technical Integration Manager
TFA	Tanks Focus Area
TISA	Technology Integration Systems Application
TMS	Technology Management System
TPO	Technical Program Officer
TPR	Technical Performance Report
TRU	Transuranic
TTP	Technical Task Plan
USG	User Steering Group
UP	University Program
USNRC	U.S. Nuclear Regulatory Commission
VOC	Volatile Organic Compound
WGA	Western Governors' Association
WIPP	Waste Isolation Pilot Plant
WM	Waste Management (commonly known by its EM mail stop, EM-30)

Appendix K

Biographical Sketches of Committee Members and Consultants

RAYMOND G. WYMER, *Chair,* is a retired director of the Chemical Technology Division of Oak Ridge National Laboratory. He is a specialist in radiochemical separations technology for radioactive waste management and nuclear fuel reprocessing. He is a consultant for the Oak Ridge National Laboratory and for DOE in the area of chemical separations technology. He is a fellow of the American Nuclear Society and the American Institute of Chemists, and has received the American Institute of Chemical Engineers' Robert E. Wilson Award in Nuclear Chemical Engineering and the American Nuclear Society's Special Award for Outstanding Work on the Nuclear Fuel Cycle. He is also a member of the Advisory Committee on Nuclear Waste for the Nuclear Regulatory Commission. He has authored many technical publications, co-authored a book on chemistry in nuclear technology, and edited another on the light-water reactor fuel cycle. He received a B.A. from Memphis State University and an M.A. and Ph.D. from Vanderbilt University.

ALLEN G. CROFF is associate director of the Chemical Technology Division at Oak Ridge National Laboratory. His areas of focus include initiation and technical management of research and development involving waste management, nuclear fuel cycles, transportation, conservation, and renewable energy. Since joining ORNL in 1974, he has been involved in numerous technical studies that have focused on waste management and nuclear fuel cycles, including (1) supervising and participating in the updating, maintenance, and implementation of the ORIGEN2 computer code; (2) developing a risk-based, generally applicable radioactive waste classification system; (3) multidisciplinary development and assessment of actinide partitioning and transmutation; and (4) leading and participating in multidisciplinary national and international technical committees. He has a B.S. in chemical engineering from Michigan State University, a nuclear engineer degree from the Massachusetts Institute of Technology, and an M.B.A. from the University of Tennessee.

MARY R. ENGLISH is a research leader of the Energy, Environment, and Resources Center at the University of Tennessee and a member of its Waste Management Research and Education Institute. She previously worked in environmental planning for state government and as a consultant. Dr. English received a B.A. in American literature from Brown University, an M.S. in regional planning from the University of Massachusetts, and a Ph.D. in sociology from the University of Tennessee. She is a member of the NRC's Board on Radioactive Waste Management.

THOMAS M. JOHNSON is currently vice president, principal hydrogeologist, and director of technical services for LFR Levine Fricke, an engineering consulting firm in California. Previously, he worked with the Illinois State Geological Survey on ground water issues. He is on the Board of Directors for the National Ground Water Association and the Association of Ground Water Scientists and Engineers. He has served on previous NRC committees, including the Panel to Evaluate State and Local Groundwater Protection Programs. He is a registered geologist, and a certified hydrogeologist, and belongs to numerous affiliated professional organizations including the American Institute of Professional Geologists. He received his B.A. in geology from Augustana College, and his M.S. in geology and water resources management from the University of Wisconsin.

DUNDAR F. KOCAOGLU is currently professor and director of the Engineering Management Program at Portland State University. Previously, he was associate professor of industrial engineering and director of the Engineering Management Program at the University of Pittsburgh for 11 years. Prior to this he served as a structural engineer with various firms. He is a member of the American Society on Engineering Education, American Society of Civil Engineers, Institute of Electrical and Electronics Engineers, Institute for Operations Research and Management Sciences, and numerous other professional organizations. Dr. Kocaoglu served on the NRC Committee to Assess Barriers and Opportunities to Improve Manufacturing at Small- and Medium-Sized Companies. He received his BSCE from Robert College (Turkey), his MSCE from Lehigh University, and his MSIE and Ph.D. in operations research and systems management from the University of Pittsburgh.

MICHAEL MENKE, president of Value Creation Associates, works primarily with research-driven companies in developing successful business and technology strategies, re-engineering their R&D management and new product development processes, and improving R&D productivity. He was a founding partner of Strategic Decisions Group and led SDG's R&D and pharmaceutical industry practices, as well as its groundbreaking benchmark study of the best decision practices of the world's leading companies. Dr. Menke has published extensively and speaks frequently on a wide range of business and innovation management topics. His consulting assignments include new product commercialization strategies, product sales forecasting and capacity planning, R&D portfolio management, and evaluation of new high-technology products and processes in a wide range of industries, including biotechnology, chemicals, medical devices, and pharmaceuticals. He has a B.A. in physics from Princeton, a M.Sc. in applied math from Cambridge, and a Ph. D. in physics from Stanford University.

GEORGE L. NEMHAUSER is currently an institute professor and holder of the Chandler Chair in the School of Industrial and Systems Engineering at the Georgia Institute of Technology. Previously, he taught at Johns Hopkins University and Cornell University. He is a member of the National Academy of Engineering. He is a past president of INFORMS (formerly the Operations Research Society of America) and past chairman of the Mathematical Programming Society. He is the founding editor of *Operations Research Letters* and former editor in chief of the JORSA. He received his B.Ch.E. from City College of New York, and his M.S. and Ph.D. from Northwestern University. He was elected to the National Academy of Engineering in 1986.

LINDA S. WENNERBERG has more than 20 years' experience in the review and development of environmental and economic management policies with technical applications. Dr. Wennerberg's diverse experience includes fostering linkages to promote use of environmental data systems by an array of decision makers, environmental assessments for DOE facilities, analysis of diversification potential and technological innovation, a performance review of a major federal toxics program, implementation of radioactive and hazardous waste management programs for state agencies, drafting siting criteria for low-level radioactive waste disposal, determination of environmental enforcement priorities for extraction industries, and identification of pollution prevention opportunities in manufacturing operations. Recent experience emphasizes the development of innovative environmental management and data systems for industrial clients to achieve compliance and competitive advantage. Dr. Wennerberg is a principal partner in Environmental Business Strategies, an environmental consulting practice based in Cambridge, Massachusetts. She received a B.S. in ecological studies and her M.S. and Ph.D. in environmental law and resource economics from Michigan State University.

EDWIN L. ZEBROSKI provides consulting services on risk analysis and decision analysis through Elgis Consulting Company. He has extensive experience in the design, development, safety, materials, fuel cycle, and economic aspects of power systems. He has more than 150 technical publications, mostly on energy-related topics, including patents and sections of seven books. Previous positions include manager of development engineering, General Electric Company; director of the Systems and Materials Department and chief nuclear scientist at the Electric Power Research Institute; vice president of engineering at the Institute for Nuclear Power Operation; and director of Risk Management Services at Aptech Engineering Co. Degrees include a B.S. in physics and chemistry from the University of Chicago and a Ph.D. in physical chemistry from the University of California, Berkeley. He was elected to the National Academy of Engineering in 1981.

CONSULTANTS

THOMAS A. COTTON is vice president of JK Research Associates, Inc., where he is a principal in activities related to radioactive waste management policy and strategic planning. Before joining JK Research Associates, he dealt with energy policy and radioactive waste management issues as an analyst and project director during nearly 11 years with the congressional Office of Technology Assessment. His expertise is in public policy analysis, nuclear waste management, and strategic planning. He received a B.S. in electrical engineering from Stanford University, a M.S. in philosophy, politics, and economics from Oxford University, and a Ph.D in engineering economic systems from Stanford University.

ROBERT J. GIORDANO is a private consultant in the areas of nuclear plant radiological and maintenance operations and decommissioning. He has 30 years of hands-on experience with DOE, the Department of Defense, and commercial nuclear power plants, both national and international. This includes the decommissioning of the Shippingport plant, where he was the manager of radiological engineering. He has served on several technical groups including the Boiling Water Reactor Owners Group, the Brookhaven National Laboratory ALARA Center, and the Nuclear Energy Institute. He has a B.S. in electrical engineering from Bucknell University.

DETLOF VON WINTERFELDT is a professor of public policy and management at the University of Southern California and director of its Institute for Civic Enterprise. He also is the president of Decision Insights, Inc., a management consulting firm specializing in decision and risk analysis. His research interests are in the foundation and practice of decision and risk analysis as applied to technology and environmental management problems. He is the coauthor of two books and author or coauthor of more than 100 articles and reports on these topics. He has served on several committees and panels of the National Science Foundation (NSF) and the National Research Council, including the NSF's Advisory Panel for its Decision and Risk Management Science Program and the NRC's Committee on Risk Perception and Risk Communication.

Appendix L

Other Materials and References Received by the Committee

The references listed in this appendix are organized into the following nine categories:

General Decision-Making References,
Non-DOE References Relevant to DOE-EM.
General DOE-EM References,
General OST References,
Crosscutting Programs References,
D&D Focus Area and Other FETC References,
Mixed Waste Focus Area and Other INEEL References,
Subcon Focus Area and Other Savannah River Site References, and
Tanks Focus Area and Other Hanford Site References.

General Decision-Making References

Arkes, Hal R., and Kenneth R., Hammond, eds. 1986. Judgment and Decision Making: An Interdisciplinary Reader. New York, NY: Cambridge University Press. (Arkes and Hammond, 1986).

Baird, B. 1989. Managerial Decisions Under Uncertainty. New York: John Wiley & Sons. (Baird, 1989).

Behn, R. D. and J. D. Vaupel. 1982. Quick Analysis for Busy Decision Makers. New York: Basic Books. (Behn, and Vaupel, 1982).

Bodily, S.E. 1985. Modern Decision Making. New York: McGraw-Hill. (Bodily, 1985).

Brown, R.V. A.S. Kahr., and C. Peterson. 1974. Decision Analysis for the Manager. New York: Holt. (Brown, Kahr, and Peterson, 1974).

Buede, Dennis M., and Daniel T. Maxwell. 1995. Rank Disagreement: A Comparison of Multi-Criteria Methodologies. Journal of Multi-Criteria Decision Analysis, Vol. 4.: John Wiley & Sons, Ltd. Pp. 1-21. (Buede and Maxwell, 1995).

Bunn, D. 1984. Applied Decision Analysis. New York: McGraw-Hill. (Buhn, 1984).

Chernoff, H. and L. Moses. 1959. Elementary Decision Theory. New York: John Wiley & Sons. (Chernoff, and Moses, 1958).

Corbin, Ruth, and A. A. J. Marley. 1974. Random Utility Models with Equality: An Apparent, but Not Actual, Generalization of Random Utility Models. Journal of Mathematical Psychology, Vol. 11: p.p. 274-293. (Corbin and Marley, 1974).

Cooper, Robert G. 1993. Winning at New Products: Accelerating the Process from Idea to Launch. Reading, MA: Addison-Wesley. (Cooper, 1993).

Dawes, R. 1988. Rational Choice in an Uncertain World. San Diego, CA.: Harcourt Brace. (Dawes, 1988).

Easton, A. 1973. Complex Managerial Decisions Involving Multiple Objectives. New York: John Wiley & Sons. (Easton, 1973).

Edwards, Allen Louis. 1954. Statistical Methods for the Behavioral Sciences. New York: Rinehart. (Edwards, 1954).

Epstein, Marc J. 1995. Measuring Corporate Environmental Performance. (Epstein, 1995).

Farquhar, Peter H., and Anthony R. Pratkanis. 1993. Decision Structuring with Phantom Alternatives. Management Sciences, Vol. 39 (10): October. Institute of Management Sciences. Pp. 1214-1226. (Farquhar and Pratkanis, 1993).

Fishburn, P.C. 1964. Decision and Value Theory. New York: John Wiley & Sons. (Fishburn, 1964).

French, S. 1986 Decision Theory: An Introduction to the Mathematics of Rationality. London: John Wiley & Sons. (French, 1986).

Hammond, John S., Ralph L. Keeney, and Howard Raiffa. 1999. Smart Choices: A Practical Guide to Making Better Decisions. Boston, MA: Harvard Business School Press. (Hammond, Keeney, and Raiffa, 1999).

Hammond, Kenneth R. 1996. Human Judgment and Social Policy: Irreducible Uncertainty, Inevitable Error, Unavoidable Injustice. New York, NY: Oxford University Press. (Hammond, 1996).

Hammond, Kenneth R., Gary H. McClelland, and Jeryl Mumpower. 1980. Human Judgment and Decision Making: Theories, Methods, and Procedures. New York, NY: Praeger Publishers. (Hammond, McClelland, and Mumpower, 1980).

Hogarth, R. 1987. Judgment and Choice, 2nd. ed . New York: John Wiley & Sons. (Hogarth, 1987).

Holloway, C.A. 1979 Decision Making Under Uncertainty. Englewood Cliffs, N.J.: Prentice Hall. (Holloway, 1979).

Howard, Ronald A. 1966. Dynamic Programming and Markov Processes. Cambridge, MA: Technology Press of MIT. (Howard, 1966).

Howard, R.A., and J.E. Matheson (eds.). 1983. The Principles and Applications of Decision Analysis. Palo Alto, CA.: Strategic Decisions Group. (Howard and Matheson, 1983).

Kahneman, Daniel, Paul Slovic, and Amos Tversky. 1982. Judgement Under Uncertainty: Heuristics and Biases. New York, NY: Cambridge University Press. (Kahneman, Slovic, and Tversky, 1982).

Keeney, Ralph L. Value-Focused Thinking: A Path to Creative Decision-Making. Cambridge, MA: Harvard University Press. (Keeney, 1992).

Keeney, R.L., and H. Raiffa. 1976. Decisions with Multiple Objectives. New York: John Wiley & Sons. (Keeney and Raiffa, 1976).

Lander, Lynn, David Matheson, Michael M. Menke, and Derek L. Ransley. 1995. Improving the R&D Decision Process. Industrial Research Institute: Research-Technology Management, January-February. . (Lander, Matheson, Menke and Ransley, 1995).

Levary, Reuven, and Neil E. Seitz. 1990. Quantitative Methods for Capital Budgeting. Cincinnati, OH: South-Western Publishing Company. (Levary and Seitz, 1990).

Lindley, D.V. 1985. Making Decisions. New York: John Wiley & Sons. (Lindley, 1985).

Luce, R.D., and H. Raiffa. 1958. Games and Decisions. New York: John Wiley & Sons. (Luce and Raiffa, 1958).

March, James G. 1994. A Primer on Decision Making: How Decisions Happen. New York, NY: The Free Press. (March, 1994).

March, James G. 1999. The Pursuit of Organizational Intelligence. Malden, MA: Blackwell Publishers, Inc. (March, 1999).

March, James G., and Roger Weissinger-Baylon. 1986. Ambiguity and Command: Organizational Perspectives on Military Decision-Making. Marshfield, MA: Pitman Publishing, Inc. (March and Weissinger-Baylon, 1986).

Martino, J. 1995. R&D Project Selection. New York: John Wiley & Sons. (Martino, 1995).

Matheson, D. and J. Matheson. 1998. The Smart Organization. Cambridge, MA: Harvard Business School Press. (Matheson and Matheson, 1998).

Matheson, David, James E. Matheson, and Michael M. Menke. 1995. Improving your R&D decision making. CHEMTECH (June 25). American Chemical Society. pp. 13-16. (Matheson, Matheson, & Menke, 1995).

Matheson, David, James E. Matheson, and Michael M. Menke. 1994. Making Excellent R&D Decisions. Industrial Research Institute: Research-Technology Management, November-December. (Matheson, Matheson, and Menke, 1994).

Matheson, James E. and Michael M. Menke. 1994. Using Decision Quality Principles to Balance Your R&D Portfolio. Industrial Research Institute: Research-Technology Management, May-June. (Matheson and Menke, 1994).

Menke, Michael M. 1997. Essentials of R&D Strategic Excellence. Research Technology Management. (Menke, 1997a).

Menke, Michael M. 1997. Managing R&D for Competitive Advantage. Research Technology Management. (Menke, 1997b).

Menke, Michael M. 1994. R&D Quality Tools. Research Technology Management. (Menke, 1994).

Pratt, J., H. Raiffa, and R. Schlaifer. 1965. Introduction to Statistical Decision Theory. New York: McGraw-Hill. (Pratt, Raiffa, and Schaifler, 1965).

Raiffa, H. 1968. Decision Analysis. Redding, Mass.: Addison-Wesley. (Raiffa, 1968).

Russo, J. Edward and Paul Schoemaker. 1989. Decision Traps: Ten Barriers to Excellent Decision Making and How to Overcome Them. New York, NY: Simon & Schuster. (Russo and Schoemaker, 1989).

Saaty, T.L. 1980. The Analytic Hierarchy Process. New York: McGraw-Hill. (Saaty, 1980).

Saaty, T.L. 1982. Decision Making for Leaders. Belmont, CA.: Lifetime Learning Publications. (Satty, 1982).

Saaty, Thomas L. 1996. Decision Making with Dependence Feedback: The Analytic Network Process. Pittsburgh, PA: RWS Publications. (Saaty, 1996).

Saaty, T.L. 1996. Multicriteria Decision Making: The Analytic Hierarchy Process. AHP Series. Pittsburgh, PA: RWS Publications. (Saaty, 1996).

Saaty, Thomas L. 1994. Fundamental of Decision Making and Priority Theory: With the Analytic Hierarchy Process (Vol. VI). Pittsburgh, PA: RWS Publications. (Saaty, 1994).

Saaty, Thomas L., and Luis G. Vargas. 1998. Diagnosis with Dependent Symptoms: Bayes Theorem and The Analytic Hierarchy Process. Operations Research, Vol. 46 (4): July-August. INFORMS. pp. 491-502. (Saaty and Vargas, 1998).

Saaty, Thomas L., and Luis G. Vargas. 1993. Experiments on Rank Preservation and Reversal in Relative Measurement. Mathematical and Computer Modeling, Vol. 17 (4/5): Pergamon Press, Ltd. Pp. 13-18. (Saaty and Vargas, 1993).

Samson, D. 1988. Managerial Decision Analysis. Homewood, IL.: Irwin. (Samson, 1988).

Schlaifer, R. 1959. Analysis of Decisions Under Uncertainty. New York: McGraw-Hill. (Schlafer, 1959).

Scott, James C. 1998. Seeing Like A State: How Certain Schemes to Improve the Human Condition Have Failed. New Haven, CT: Yale University Press. (Scott, 1998).

Simon, Herbert A. 1997. Administrative Behavior: A Study of Decision-Making Processes in Administrative Organizations (fourth edition). New York, NY: The Free Press. (Simon, 1997).

Simon, Herbert A. 1977. The New Science of Management Decision (revised edition). Englewood Cliffs, NJ: Prentice-Hall, Inc. (Simon, 1977).

Simon, Herbert A. 1996. The Sciences of the Artificial (third edition). Cambridge, MA: The MIT Press. (Simon, 1996).

Simon, Herbert A., Donald W. Smithburg, and Victor A. Thompson. 1991. Public Administration. New Brunswick, NJ: Transaction Publishers. (Simon, Smithburg, and Thompson 1991).

Sjöberg, Lennart, Tadeusz Tyszka, and James A. Wise, eds. 1983. Human Decision Making. Sweden: Lagerblads Tryckeri AB. (Sjöberg, Tyszka, and Wise, 1983).

Slowinski, Gene, Deb Chatterjai, James A. Tshudy, and Dale L. Fridley. 1997. Are you A leader in Environmental R&D? Research Technology Management, Vol. 40 (3): Washington, D.C.: Industrial Research Institute, Inc. May-June. (Slowinski et al., 1997).

Stigum, Bernt P., and Fred Wenstop. 1983. Foundations of Utility and Risk Theory with Applications. Dordrecht, Holland: D. Reidel Publishing Company. (Stigum and Wenstop, 1983).

Tribus, Myron. 1969. Rational Descriptions, Decisions, and Designs. New York: Pergamon Press. (Tribus, 1969).

Tuchman, Barbara. 1992. The March of Folly: From Troy to Vietnam. Ballantine Books, NY. (Tuchman, 1992).

Tversky, Amos, and Daniel Kahneman. 1974. Judgment and Uncertainty: Heuristics and Biases. Science. Vol. 185: September. pp. 1124-1131. (Tversky and Kahneman, 1974).

Tversky, Amos, and Itamar Simonson. 1993. Context-Dependent Preferences. Management Science. Vol. 39 (10): October. Institute of Management Sciences. pp. 1179-1189. (Tversky and Simonson, 1993).

Tversky, Amos, Paul Slovic, and Daniel Kahneman. 1990. The Causes of Preference Reversal. The American Economic Review. Vol. 80 (1): March. pp. 204-217. (Tversky, Slovic, and Kahneman, 1990).

Von Neumann, J., and O. Morgenstern. 1953. Theory of Games and Economic Behavior. Revised edition. Princeton, N.J.: Princeton University Press. (Von Neumann and Morgenstern, 1953).

Von Winterfeldt, D., and W. Edwards. 1986. Decision Analysis and Behavioral Research. Cambridge: Cambridge University Press. (Von Winterfeldt and Edwards, 1986).

Watson, S., and D. Buede. 1987. Decision Synthesis. Cambridge: Cambridge University Press. (Watson and Buede, 1987).

Winkler, R.L. 1972. Introduction to Bayesian Inference and Decision. New York: Holt, Rinehart and Winston. (Winkler, 1972).

Yarmolinsky, Adam. 1987. Governance of the U.S. Military Establishment. University Press of America. (Yarmolinsky, 1987).

Zeleny, Milan. 1982. Multiple Criteria Decision Making. New York, NY: McGraw-Hill Book Company. (Zeleny, 1982).

Non-DOE References Relevant to DOE-EM

American Society of Mechanical Engineers. 1997. Assessment of Technologies supported by the U.S. Department of Energy Office of Science and Technology: Results for the Peer Review for Fiscal Year 1997. American Society of Mechanical Engineers. (ASME, 1997a).

American Society of Mechanical Engineers. 1997. Peer Review of MAG*SEP[SM]. Review on April 23-25, in Atlanta, GA. (ASME, 1997b).

Berkey, Dr. Edgar. 1997. EMAB Perspective on OST Decision Making. Presentation to the Committee on November 10. (Berkey, 1997).

Berkey, Dr. Edgar. 1998. Review of Focus Area Program. Paper presented to Department of Energy, Environmental Management Advisory Board on January 21. (Berkey, 1998).

Blush, Stephen M. and Thomas H. Heitman. 1995. Train Wreck Along the River of Money: An Evaluation of the Hanford Cleanup. Report presented to United States Senate Committee on Energy and Natural Resources. March. (Blush & Heitman, 1995).

Bournakis, A.D., and G.D. Pine. 1997. Benefits of GRI R&D Results that Have Been Placed in Commercial Use in 1992 through 1996. Report No. GRI 97/0164. Chicago, IL: University of Illinois at Chicago Energy Resources Center. (Bournakis and Pine, 1997).

Burns, Conrad, Max Baucus, and Rick Hill. 1997. Correspondence to Federico Pena regarding WETO budget reductions. February 27. (Burns, Baucus & Hill, 1997).

E.I. Du Pont De Nemours and Company. 1993. Corporate Environmentalism: 1993 Progress Report. DuPont Printing and Publishing. (DuPont, 1993).

Environmental Protection Agency. 1997. Completed North American Innovative Remediation Technology Demonstration Projects. (EPA, 1997).

Farber, D. 1996. Scenario Planning, the Interpretation of Uncertainty and Value-Focused Decision Making: A Case Study of a Radioactive Waste Site Remediation. Proceedings of June 21-23 meeting of International Symposium on Technology and Society Technical Expertise and Public Decisions. New York: IEEE. (Farber, 1996).

Gas Research Institute. 1996. Gas Research Institute: 1997-2001 Research and Development Plan and 1997 Research and Development Program. Chicago, IL: GRI. April. (GRI, 1996a).

Gas Research Institute. 1996. Results of Appraisal of GRI 1997-2001 R&D Plan. Chicago, IL: GRI. May.

Gas Research Institute. 1997. Basic Research Results (1996 results). Chicago, IL: GRI. June. (GRI, 1997a).

Gas Research Institute. 1997. Gas Research Institute: 1998-2002 Research and Development Plan and 1998 Research and Development Program. Chicago, IL: GRI. April. (GRI, 1997b).

Gas Research Institute. 1997. GRID: Gas Research Institute Digest/Volume 19 (4)., Winter 1996/1997. Chicago, IL: GRI.

Gas Research Institute. 1997. Results of Appraisal of GRI 1998-2002 R&D Plan. Chicago, IL: GRI. June. (GRI, 1997c).

Gas Research Institute. 1998. Gas Research Institute 1999 to 2003 Plan and 1999 Research and Development Program. Chicago, IL: GRI. (GRI, 1998).

Gas Research Institute, and Planning Research Corporation. 1997. Defining the Value of Technology. Summary of Pipeline Industry Technology Workshop, May 12-14 in Denver, CO. (GRI, 1997).

General Accounting Office. 1992. Cleanup Technologies. Washington, D.C.: General Accounting Office. (GAO, 1992).

General Accounting Office. 1994. Management Changes Needed to Expand Use of Innovative Cleanup Technologies: Report to the Secretary of Energy. Washington, D.C.: General Accounting Office. August. (GAO, 1994).

General Accounting Office. 1996. Energy Management: Technology Development Program Taking Action to Address Problems. Washington, D.C.: General Accounting Office. July. (GAO, 1996).

General Accounting Office. 1998. Nuclear Waste: Department of Energy's Hanford Tank Waste. letter Report. Washington, D.C.: General Accounting Office. October 8. (GAO, 1998).

General Accounting Office. 1998. Nuclear Waste: Further Actions Needed to Increase the Use of Innovative Cleanup Technologies. Report to Congressional Committees. Washington, D.C.: General Accounting Office. September.

Holmes, J.D. 1994. White Water Ahead: Eastman Prepares for Turbulent Times. Research Technology Management. (Holmes, 1994).

Holt, Mark and Jeff Day. 1997. Memorandum, Congressional Research Service: Mark Holt & Jeff Day to Senate Appropriations Committee re Environmental Technology Development at DOE. Congressional Research Service. (Holt and Day, 1997).

Interviews with GRI Managers, September 4-5, 1997, Chicago, IL: Gas Research Institute.

Martin, H. Dale. 1997. Environmental Planning--Balancing Environmental Commitments with Economic Realities. Speech before the Southeastern Electrical Exchange, Charlotte, NC. June 26. (Martin, 1997).

Martin, H. Dale. 1994. Environmental Planning: Balancing Environmental Commitments with Economic Realities. Presentation to the Proceedings of the Environmental Management in a Global Economy (GEMI'94) Conference. (Martin, 1994).

Menke, Michael M. 1995. Lessons Learned from Industrial R&D Management, Strategic Decisions Group. presented to the Committee on Environmental Management Technologies and the National Research Council. (Menke, 1995).

National Environmental Technology Applications Center. 1995. Barriers to Environmental Technology Commercialization. Pittsburgh, PA: University of Pittsburgh Applied Research Center. April. (NETAC, 1995).

National Research Council. 1989. Nuclear Weapons Complex: Management for Health, Safety, and the Environment. Washington, D.C.: National Academy Press. (NRC, 1989).

National Research Council. 1994. Building Consensus Through Risk Assessment and Management of the Department of Energy's Environmental Remediation Program. Washington, D.C.: National Academy Press. (NRC, 1994).

National Research Council. 1995. The Committee on Environmental Management Technologies Report for the Period Ending December 31, 1994. Washington, D.C.: National Academy Press. (NRC, 1995b).

National Research Council. 1995. Improving the Environment. Washington, D.C.: National Academy Press. (NRC, 1995a).

National Research Council. 1995. Research Restructuring and Assessment: Can We Apply the Corporate Experience to Government Agencies? Washington, D.C.: National Academy Press.

National Research Council. 1996. Barriers to Science: Technical Management in the Department of Energy Environmental Remediation Program. Washington, D.C.: National Academy Press. (NRC, 1996a).

National Research Council. 1996. Environmental Management Technology-Development Program at the Department of Energy 1995 Review. Washington, D.C.: National Academy Press. (NRC, 1996b).

National Research Council. 1996. The Hanford Tanks: Environmental Impacts and Policy Choices. Washington, D.C.: National Academy Press. (NRC, 1996c).

National Research Council. 1996. Understanding Risk: Informing Decisions in a Democratic Society. Washington, D.C.: National Academy Press. (NRC, 1996d).

National Research Council. 1997. Building an Effective Environmental Management Science Program: Final Assessment. Washington, D.C.: National Academy Press. (NRC, 1997a).

National Research Council. 1997. Building a Foundation for Sound Environmental Decisions. Washington, D.C.: National Academy Press.

National Research Council. 1997. Enhancing Organizational Performance. Washington, D.C.: National Academy Press.

National Research Council. 1997. Peer Review in the Department of Energy-Office of Science and Technology: Interim Report. Washington, D.C.: National Academy Press. (NRC, 1997b).

National Research Council. 1998. A Review of Decontamination and Decommissioning Technology Development Programs at the Department of Energy. Washington, D.C.: National Academy Press. (NRC, 1998a).

National Research Council. 1998. Assessing the Need for Independent Project Reviews in the Department of Energy. Washington, D.C.: National Academy Press.

National Research Council. 1998. Peer Review in Environmental Technology Development Programs: The Department of Energy--Office of Science and Technology. Washington, D.C.: National Academy Press. (NRC, 1998b).

National Research Council. 1999. An End State Methodology for Identifying Technology Needs for Environmental Management: With an Example from the Hanford Site Tanks. National Academy Press, Washington, D.C. (NRC, 1999a)

National Research Council. 1999. Evaluating Federal Research Programs: Research and the Government Performance and Results Act. Washington, D.C.: National Academy Press. (NRC, 1999b).

National Transportation Safety Board. 1995. Pipeline Accident Report: Texas Eastern Transmission Corporation Natural Gas Pipeline Explosion and Fire, Edison, New Jersey, March 23, 1994. Washington, D.C.: National Transportation Safety Board. (NTSB, 1995).

Perdue, R.K., B.G. Berkey, P.V. King, and W.J. McAllister. 1997. Valuation of R&D Projects Using Options Pricing & Decision Analysis Models--Theory and Application. Final Draft submitted to

Management Sciences on March 20. Pittsburgh, PA: Westinghouse Science and Technology Center. (Perdue, Berkey, King, and McAllister, 1997).

Rezendes, Victor S. 1997. Cleanup Technology: DOE'S Program to Develop New Technologies for Environmental Cleanup. Testimony before the Unites States House of Representatives Subcommittee on Oversight and Investigations. Washington, D.C.: General Accounting Office. (Rezendes, 1997).

Russell, Milton. 1998. Toward a Productive Divorce: Separating DOE Cleanups from Transition Assistance. Knoxville, TN: The Joint Institute for Energy and Environment. (Russell, 1998).

Russell, Milton, E. William Colglazier, and Mary English. 1991. Hazardous Waste Remediation: The Task Ahead. Knoxville, TN: Waste Management Research and Education Institute, University of Tennessee. December. (Russell, Colglazier and English, 1991).

Surles, Terry. 1997. Correspondence to Al Alm with viewgraphs prepared by Strategic Laboratory Council on March 7 entitled "Review Recommendations for Office of Science and Technology Policy". May 15. (Surles, 1997).

United States Army Corps of Engineers. 1997. Project EM Task Force Phase 1 Report. Washington, D.C.: U.S. Army Corps of Engineers. February. (U.S. Army Corps of Engineers, 1997).

United States Congress, Office of Technology Assessment. 1986. Research Funding as an Investment: Can We Measure the Returns?--A Technical Memorandum. Washington, D.C.: U.S. Congress/Government Printing Office. April. (U.S. Congress, 1986).

United States Congress. 1989. National Defense Authorization Act for Fiscal Years 1990 and 1991. P.L. 101-189, Sec. 3141, (H.R. 2461). Washington, D.C.: Government Printing Office. (U.S. Congress, 1989).

United States Congress, Senate. 1992. Energy and Water Development Appropriations Bill, 1993. Report to accompany H.R. 5373. 102d Congress, 2d session, pp. 138-139. July 27. (U.S. Congress, 1992).

United States Congress, House of Representatives. 1994. Energy and Water Development Appropriations Bill, 1995. Report to accompany H.R. 4506. 103d Congress, 2d session, pp.76-77, 100. May 26. (U.S. Congress, 1994).

United States Congress, House of Representatives. 1995. Energy and Water Development Appropriations Bill, 1996. Report to accompany H.R. 1905. 104th Congress, 1st session, pp. 77-78. June 20. (U.S. Congress, 1995a).

United States Congress, House of Representatives. 1995. Making Appropriations for Energy and Water Development for the Fiscal Year Ending September 30, 1996, and for Other Purposes. Conference Report to accompany H.R.1905. 104th Congress, 1st session, pp. 69-70. October 26. (U.S. Congress, 1995).

United States Congress, Office of Technology Assessment. 1995. Innovation and Commercialization of Emerging Technologies. Washington, D.C.: Government Printing Office. September.

United States Congress, Senate. 1995. Energy and Water Development Appropriations Bill, 1996. Report to accompany H.R. 1905. 104th Congress, 1st session, pp. 110-111. July 27. (U.S. Congress, 1995b).

United States Congress, House of Representatives, Committee on Commerce, Subcommittee on Oversight and Investigations. 1997. Summary of Congressional Testimony on Department of Energy/Office of Science and Technology. Washington, D.C.: Committee on Commerce. May 7. (U.S. Congress, 1997).

Willke, Theodore L., Theresa M. Shires, R. Michael Cowgill, and Bernd J. Selig. 1997. U.S. Risk Management Can Reduce Regulation, Enhance Safety. *Oil & Gas Journal* (June 16). Pennwell Publishing Company. (Willke et al., 1997).

General DOE-EM References

Alm, Alvin. 1996. Accelerated Environmental Cleanup. Office of Environmental Management, United States Department of Energy. (Alm, 1996).

Alm, Alvin. 1998. Action Items and Understandings form the Technology Accleration Committee Meeting of January 27, 1998. Department of Energy. Memorandum for Distribution. January 30. (Alm, 1998).

Alm, Alvin. 1997. Technology Deployment. Department of Energy. Memorandum for Distribution. July 3. (Alm, 1997).

Barainca, Michael. 1997. Technology Acceleration Committee Meeting Draft Notes. Department of Energy. September 9. (Barainca, 1997).

Department of Energy. 1992. Annual Report to Congress FY 1992. Washington, D.C.: Department of Energy, Office of Environmental Restoration and Waste Management. (DOE, 1992).

Department of Energy. 1993. A New Approach to Environmental Research and Technology. Development at DOE: Report of the Working Group (Viewgraphs). Washington, D.C.: Department of Energy. December 14. (DOE, 1993a).

Department of Energy. 1993. Annual Report to Congress FY 1993. Washington, D.C.: Department of Energy, Office of Environmental Management. (DOE, 1993).

Department of Energy. 1993. EM Progress, Summer 1993. Washington, D.C.: Department of Energy. (DOE, 1993b).

Department of Energy. 1993. A Strategy for Environmental Technology and Economic Competitiveness: Commitment for Change. Washington, D.C.: Department of Energy, Strategic Laboratory Council. January 20. (DOE, 1993).

Department of Energy. 1994. A New Approach to Environmental Research and Technology Development at the U.S. Department of Energy: Action Plan. Washington, D.C.: Department of Energy. January 25. (DOE, 1994a).

Department of Energy. 1994. Annual Report to Congress FY 1994 (Predecisional Draft). Washington, D.C.: Department of Energy, Office of Environmental Management. (DOE, 1994).

Department of Energy. 1994. Committed to Results: DOE's Environmental Management Program, An Introduction. Washington, D.C.: Department of Energy. April. (DOE, 1994).

Department of Energy. 1994. Fueling a Competitive Economy: Strategic Plan. Washington, D.C.: Department of Energy. April. (DOE, 1994).

Department of Energy. 1994. FY 1994 Program Summary. Washington, D.C.: Department of Energy. October. (DOE, 1994).

Department of Energy. 1994. FY 1995 Technology Development Needs Summary. Washington, D.C.: Department of Energy, Office of Environmental Management. March. (DOE, 1994b).

Department of Energy. 1994. Highest Risk Sites in the Department of Energy Environmental Restoration Program (Draft). Washington, D.C.: Department of Energy. April 11. (DOE, 1994).

Department of Energy. 1994. Technology Management Process Document (Predecisional Draft). April. Washington, D.C.: Department of Energy. April 25. (DOE, 1994).

Department of Energy. 1995. Annual Report to Congress, FY 1995. Washington, D.C.: Department of Energy, Office of Environmental Management. (DOE, 1995).

Department of Energy. 1995. Estimating the Cold War Mortgage: The 1995 Baseline Environmental Management Report, Vol. 1-3. Washington, D.C.: Department of Energy, Office of Environmental Management. March. (DOE, 1995c).

Department of Energy. 1995. Guidance for FY 1998 Budget Formulation, Part I. Washington, D.C.: Department of Energy. November. (DOE, 1995).

Department of Energy. 1995. OST Background Series: 2: Final Report of the Task Force On Alternative Futures for the Department of Energy National Laboratories (Galvin Report). Washington, D.C.: Department of Energy. February 1. (DOE, 1995e).

Department of Energy. 1995. Site Needs Assessment Report. Washington, D.C.: Department of Energy. (DOE, 1995g).

Department of Energy. 1996. Annual Report to Congress FY 1996. Washington, D.C.: Department of Energy, Office of Environmental Management. (DOE, 1996).

Department of Energy. 1996. Environmental Management 1996: Progress and Plans of the Environmental Management Program. Washington, D.C.: Department of Energy. (DOE, 1996).

Department of Energy. 1996. FY 1997 and 1998 Annual Performance Plan Handbook. Washington, D.C.: Department of Energy. September 13. (DOE, 1996).

Department of Energy. 1996. Strategic Plan 1996-2005. Washington, D.C.: Department of Energy, Office of Environmental Management. June. (DOE, 1996).

Department of Energy. 1997. Accelerating Cleanup: Focus on 2006 (Discussional Draft). Department of Energy, Office of Environmental Management. June. (DOE, 1997b).

Department of Energy. 1997. Focus Area Technology Investments Respond to Ten Year Plan. Washington, D.C.: Department of Energy. (DOE, 1997).

Department of Energy. 1997. FY 1998 Budget Overview (viewgraphs). Washington, D.C.: Department of Energy. February 25. (DOE, 1997).

Department of Energy. 1997. Linking Legacies: Connecting the Cold War Nuclear Weapons Production Processes to Their Environmental Consequences. Washington, D.C.: Department of Energy. January. (DOE, 1997).

Department of Energy. 1997. Strategic Plan. Washington, D.C.: Department of Energy. September. (DOE, 1997j).

Department of Energy. 1997. Technology Management Review. Washington, D.C.: Department of Energy, Office of Science and Technology. July 8. (DOE, 1997k).

Department of Energy. 1998. Accelerating Cleanup: Paths to Closure. Washington, D.C.: Department of Energy/Office of Environmental Management. June. (DOE, 1998a).

Department of Energy. 1998. Accelerating Cleanup: Paths to Closure (Draft). Washington, D.C.: Department of Energy. February. (DOE, 1998b).

Department of Energy. 1998. Selection Processes for Science and Technology Projects. Washington, D.C.: Department of Energy, Strategic Laboratory Council. (DOE, 1998).

Department of Energy. 1998. White Paper on DOE Environmental Management Science Program. Washington, D.C.: Department of Energy, Strategic Laboratory Council. April 6. (DOE, 1998).

O'Leary, Hazel R. 1995. Statement of Energy Secretary Hazel R. O'Leary in response to the report to the U.S. Senate Committee on Energy and Natural Resources: Train Wreck Along the River of Money. Washington, D.C.: Department of Energy. March 14. (O'Leary, 1995).

General OST References

Barainca, Michael. 1998. New OST Process for Prioritization: Using the Accelerating Cleanup: Paths to Closure Data to make Budget Decisions. Paper Presented to Committee on April 2. (Barainca, 1998).

Bauer, Carl O. 1996. Implementation of New Peer Review Policy. Washington, D.C.: Department of Energy. Memorandum for Distribution. (Bauer, 1996).

Department of Energy. 1994. Technology Management Process Document (Predecisional Draft). Washington, D.C.: Department of Energy.

Department of Energy. 1995. Technology Needs/Opportunities Statement Outline (undated compilation of early need gathering template, Rocky Flats examples). Washington, D.C.: Department of Energy.

Department of Energy. 1996. FY 1998 Internal Review Budget Preparation Handbook. Washington, D.C.: Department of Energy, Office of Science and Technology. March 14.

Department of Energy. 1996. Office of Science and Technology (OST) Environmental Technology Decision Process Revision No. 5. Washington, D.C.: Department of Energy, Office of Science and Technology. September 6.

Department of Energy. 1996. Office of Science and Technology Decision Process Standard Operating Procedures (Draft). Washington, D.C.: Department of Energy. December 11. (DOE, 1996o).

Department of Energy. 1996. PI Requirements and Deliverables for the OST Technology Decision Process Procedures. Washington, D.C.: Department of Energy. (DOE, 1996p).

Department of Energy. 1996. Research Opportunity Announcement (ROA) for Applied Research and Development of Technologies for Environmental Restoration and Waste Management. Washington, D.C.: Department of Energy.

Department of Energy. 1996. Research Opportunity Announcement (ROA) Proposal Evaluation Plan DEAR 935.016-4(d). Washington, D.C.: Department of Energy.

Department of Energy. 1996. Strategic Plan: 1996-2005; Environmental Science and Technology, (Draft).

Department of Energy. 1997. EM International, Vol. 3. Washington, D.C.: Department of Energy. Spring.

Department of Energy. 1997. EMSP Research Needs Collected for the Environmental Management Science Program Series Facility Deactivation and Decommissioning. Washington, D.C.: Department of Energy. June.

Department of Energy. 1997. EMSP Research Needs Collected for the Environmental Management Science Program Series Health/Ecology/Risk. Washington, D.C.: Department of Energy. June.

Department of Energy. 1997. EMSP Research Needs Collected for the Environmental Management Science Program Series High-Level Radioactive Waste. Washington, D.C.: Department of Energy. June.

Department of Energy. 1997. EMSP Research Needs Collected for the Environmental Management Science Program Series Mixed Wastes. Washington, D.C.: Department of Energy. June.

Department of Energy. 1997. EMSP Research Needs Collected for the Environmental Management Science Program Series Nuclear Materials. Washington, D.C.: Department of Energy. June.

Department of Energy. 1997. EMSP Research Needs Collected for the Environmental Management Science Program Series Remedial Action. Washington, D.C.: Department of Energy. June.

Department of Energy. 1997. EMSP Research Needs Collected for the Environmental Management Science Program Series: Spent Nuclear Fuel. Washington, D.C.: Department of Energy. June.

Department of Energy. 1997. Information from July 1997 PEG Meeting in Gaithersburg. Washington, D.C.: Department of Energy. (DOE, 1997f).

Department of Energy. 1997. OST Background Series: 3: COE Reports Relevant to the Office of Science and Technology. Washington, D.C.: Department of Energy. April. (DOE, 1997h).

Department of Energy. 1997. OST Process for Prioritization, (Predecisional draft). Washington, D.C.: Department of Energy. July 14. (DOE, 1997i).

Department of Energy. 1998. OST FY98 Peer Review Schedule with Delivery Dates. Washington, D.C.: Department of Energy. January 20.

Department of Energy. 1998. Status Report on the OST Peer Review Program. Paper presented to Peer Review Panel, National Academy of Sciences on April 7.

Department of Energy. undated. Technology Needs/Opportunities Statement Outline (sample). Washington, D.C.: Department of Energy.

DuCharme, A.R., R.D. Jamenez, and W.J. Roberds. 1992. International Technology Transfer to Support the Environmental Restoration Needs of the DOE Complex. Washington, D.C.: Department of Energy. (DuCharme, Jamenez, and Roberds, 1992).

Frank, C.W. 1991. Technology Strategy for the Department of Energy Office of Environmental Restoration and Waste Management. Washington, D.C.: Department of Energy. (Frank, 1991).

Frank, C.W. 1994. Strategic Plan. Washington, D.C.: Department of Energy. August 18. (Frank, 1994).

Frank, Clyde. 1997. Office of Science and Technology Update and Program Data Analysis. Presentation to Committee on June 25. (Frank, 1997a).

Frank, Clyde. 1997. Presentation to the Committee, March 20, 1997. Washington, D.C.: Department of Energy. (Frank, 1997b).

Frolio, Nancy. 1996. Technology Decision Process Pilot Program. Washington, D.C.: Department of Energy. Memorandum to James Wright and Carl Bauer. November 14. (Frolio, 1996).

Heeb, Michael. 1996. Implementation of New Peer Review Policy. Washington, D.C.: Department of Energy. Memorandum for Distribution. November 18. (Heeb, 1996).

Huffman, Gary, and Elisabeth Reber-Cox. 1996. STCG National Meeting Report, 19-20 June, 1996 Golden, Colorado. Paper presented to Committee on July 30 – August 1. (Huffman and Cox, 1996).

Lankford, Mac. 1997. Technology Integrations Systems Applications Programs--Domestic Programs--International Programs. Paper presented to the Committee on November 10, 1997. (Lankford, 1997).

Mathur, John. 1997 . Discussion with the Committee, June. 1997. Washington, D.C.: Department of Energy. (Mathur, 1997).

Paladino, Joseph. 1996. Proposed EM Priority Setting Process. Paper presented to Committee on August 30. (Paladino, 1996).

Paladino, Joseph and Brian Fox. Undated. A Framework for R&D Planning in EM (Draft). Paper prepared for the TD Council and EM Focus Group. (Paladino and Fox, undated).

Paladino, Joseph and Paul Longworth. 1995. Maximizing R&D Investments in the Department of Energy's Environmental Cleanup Program. Technology Transfer, December. pp. 96-107. (Paladino and Longworth, 1995).

Urban Energy and Transportation Corporation. 1996. Community Leaders Network Update/Vol.2 (2). November 8. (Urban Energy, 1996).

United States Department of Energy and the Russian Federation. 1997. Environmental Management Activities/US-Russia. Washington, D.C.: Department of Energy. Spring. (DOE, 1997).

Walker, Jef. 1996. Overview from a National Perspective. Briefing to Decision-Making Processes for Selection and Prioritization Panel. Washington, D.C.: Department of Energy. November 12. (Walker, 1996).

Walker, Jef. 1997. Current Status and Future Directions of Science and Technology in the EM Program. Paper presented to National Academy of Sciences Committee on Peer Review. (Walker, 1997a).

Walker, Jef. 1997. New OST Process for Prioritization: Making OST Decisions More Responsive to the DOE Sites through the 2006 Plan. Paper Presented to Committee on November 10. (Walker, 1997b).

Walker, Jef. 1997. Technology Deployment Initiative Viewgraphs. Presentation to Committee on March 20-21.

Walker, Jef. 1998. Peer Review Panel Dialogue for National Academy of Sciences held on April 7.

Walker, Jef. 1998. Status of OST Peer Review. Paper presented to Committee on February 9.

Wengle, John. 1997. Email Correspondence from to Carolyn Davis regarding NAS request for Information. October 30.

Wolf, Stanley. 1997. Email Correspondence from to Carolyn Davis regarding Reply to ORO Information Request from NAS. October 24.

Crosscutting Programs References

Department of Energy. 1995. ResonantSonic™ Drilling Innovative Technology Summary Report. Denver, CO: Colorado Center for Environmental Management. April.

Department of Energy. 1995. Seamist™ Innovative Technology Summary Report. Denver, CO: Colorado Center for Environmental Management. August.

Department of Energy. 1996. Characterization, Monitoring & Sensor Technology Crosscutting Program: Technology Summary. Washington, D.C.: Department of Energy, Office of Science and Technology. August.

Department of Energy. 1996. Cone Penetrometer Innovative Technology Summary Report. Denver, CO: Colorado Center for Environmental Management. April.

Department of Energy. 1996. Efficient Separations & Processing Crosscutting Program: Technology Summary. Washington, D.C.: Department of Energy. August.

Department of Energy. 1996. Efficient Separations and Processing Crosscutting Program Review Process. Washington, D.C.: Department of Energy. July 3. (DOE, 1996d).

Department of Energy. 1996. FY 1996 Characterization, Monitoring, and Sensor Technology Crosscutting Program Summary Report (Draft). Paper presented at The Office of Science and Technology Crosscutting Program Review, Gaithersburg, MD. June 14. Washington, D.C.: Department of Energy.

Department of Energy. 1996. Leveling the Playing Field. Washington, D.C.: Department of Energy. July 3. (DOE, 1996j).

Department of Energy. 1996. Office of Science and Technology Program Review of Characterization, Monitoring, and Sensor Technology-Crosscutting Program. Washington, D.C.: Department of Energy. June 14.

Department of Energy. 1996. Robotics Crosscutting Program: Technology Summary. Washington, D.C.: Department of Energy/Office of Science and Technology. August.

Department of Energy. 1996. Robotics Technology Development Program Major Thrusts/Major Milestones Micro TTPs Fiscal Year 1997. Albuquerque, NM: Department of Energy/Albuquerque Operations Office. (DOE, 1996t).

Department of Energy. 1996. The ESP Project Priority Procedure (Predecisional Draft). Washington, D.C.: Department of Energy.

Department of Energy. 1997. April 1997 Progress Reports: Characterization, Monitoring & Sensor Technologies. Washington, D.C.: Department of Energy, Federal Energy Technology Center.

Department of Energy. 1997. Efficient Separations and Processing Crosscutting Program's (ESP-CP). Presentation to the National Academy of Sciences/Committee for Environmental Management Technologies (NAS/CEMT) Decision Making Subcommittee on June 6. (DOE, 1997d).

Department of Energy. 1997. In Situ Permeable Flow Sensor Innovative Technology Summary Report. Aiken, SC: Department of Energy, Westinghouse Savannah River Site. April.

Department of Energy. undated. Review Procedures of the Characterization, Monitoring, and Sensor Technology Crosscutting Program (CMST-CP). Washington, D.C.: Department of Energy.

Gephart, J.M., ed. 1997. Proceedings of the Efficient Separations and Processing Crosscutting Program 1997 Technical Exchange Meeting, January 28-30. Richland, WA: Pacific Northwest National Laboratory. (Gephart, 1997).

Matalucci, Rudolph V., Charlene Esparaza-Baca, and Richard D. Jimenez. 1995. Characterization, Monitoring, and Sensor Technology Catalogue. Albuquerque, NM: Sandia National Laboratories. December. (Matalucci, Esparanza-Baca and Jimenez, 1995).

Purdy, Caroline. 1997. Decision Making Review: Characterization, Monitoring, Sensor Technology. Presentation to Committee. (Purdy, 1997).

Special Technologies Laboratory. 1997. FY 1997 STCG Needs for Characterization, Monitoring, and Sensor Technologies. Santa Barbara, CA: Special Technologies Laboratory.

Wang, Paul. 1997. Correspondence from Paul Wang to CEMT Committee Members. May 23. (Wang, 1997).

Yarbrough, Linton. 1997. Robotics Technology Development Program. Presentation to Committee on June 6. (Yarbrough, 1997).

D&D Focus Area and other FETC References

Alm, Alvin, and Modesto Maidique. 1996. Correspondence with attachment regarding X'Change 97. Washington, D.C.: Department of Energy. October 23. (Alm & Maidique, 1996).

Bechtel, Thomas F. 1995. Request for Letter Proposals--Large Scale Decontamination and Decommissioning Demonstration Projects. Morgantown, WV: Department of Energy, Federal Energy Technology Center. Memorandum for distribution. (Bechtel, 1995).

Bechtel, Thomas F. 1996. Memorandum; Request for Letter Proposals--Large Scale Decontamination and Decommissioning (D&D) Demonstration Projects (Solicitation #2). Morgantown, WV: Department of Energy, Federal Energy Technology Center. Memorandum for distribution. (Bechtel, 1996).

Bedick, Robert C., Steven J. Bossart, and Paul W. Hart. 1995. Record of the Facility Deactivation, Decommissioning, and Material Disposition (D&D) Workshop: A New Focus for Technology Development, Opportunities for Industry/Government Collaboration. Morgantown, WV: Department of Energy, Federal Energy Technology Center. (Bedick, Bossart, and Hart 1995).

Bedick, Robert C. 1997. Industry & University Programs Overview (Viewgraphs). Presentation to Committee on May 9. (Bedick, 1997).

Black, David B. 1996. CP5 LSDP: Demonstration Technology Evaluation and Selection Process. Paper presented to Committee on March 12. (Black, 1996).

Christy, C. Edward. 1997. Industry Program Implementation. Presentation to Committee on May 9. (Christy, 1997).

Delphinus Engineering. 1995. Technology Demonstration Selection Criteria [C-Reactor]. Philadelphia, PA: Delphinus Engineering. June 12. (Delphinus Engineering, 1995).

Delphinus Engineering, Inc. undated. Reactor 105-C Interim Storage Project Technology Selection Meeting Process Overview. Philadelphia, PA: Delphinus Engineering. (Delphinus Engineering, undated).

Department of Energy. 1994. Technical Program Plan for the Transitioning, Decontamination, and Final Disposition Focus Area. Washington, D.C.: Department of Energy. January. (DOE, 1994).

Department of Energy. 1995. Evaluation Plan: Large-Scale Decontamination and Decommissioning Demonstration Projects. Washington, D.C.: Department of Energy. August 17. (DOE, 1995).

Department of Energy. 1995. Source Selection Statement for Large-Scale Decontamination and Decommissioning (D&D) Demonstration Projects (Solicitation #1). Washington, D.C.: Department of Energy. July 21. (DOE, 1995).

Department of Energy. 1995. Technology Development Through Industrial Partnerships. Morgantown, WV: Department of Energy, Federal Energy Technology Center. (DOE, 1995).

Department of Energy. 1996. 105-C Reactor LSDP IC Team Meeting Agenda. Richland, WA: Department of Energy. July 23. (DOE, 1996).

Department of Energy. 1996. Agenda RL and ANL Visit to Fernald Plant 1 Large Scale Technology Demonstration. Washington, D.C: Department of Energy. March 13. (DOE, 1996).

Department of Energy. 1996. Decontamination and Decommissioning Focus Area Annual Report 1996. Washington, D.C.: Department of Energy. (DOE, 1996c).

Department of Energy. 1996. Decontamination and Decommissioning Focus Area Quarterly Report, October-December 1996 Activity. Morgantown, WV: Department of Energy, Federal Energy Technology Center. (DOE, 1996).

Department of Energy. 1996. Decontamination and Decommissioning Focus Area Technology Summary, August 1996. Washington, D.C.: Department of Energy. (DOE, 1996).

Department of Energy. 1996. Decontamination and Decommissioning Focus Area Results of the National Needs Assessment, July 24-25, 1996. Washington, D.C.: Department of Energy. (DOE, 1996aa).

Department of Energy. 1996. Evaluation Plan: Large-Scale Demonstration Projects, Decontamination and Decommissioning Focus Area. Washington, D.C.: Department of Energy. May 30. (DOE, 1996f).

Department of Energy. 1996. Pipe Explorer (tm) System: Innovative Technology Summary Report. Oak Ridge, TN: Hazardous Waste Remedial Actions Program. April. (DOE, 1996q).

Department of Energy. 1996. Technology Needs Statements. Washington, D.C.: Department of Energy. (DOE, 1996x).

Department of Energy. 1996. Technology Screening and Evaluation--Fernald Plant 1 Case Histories. Washington, D.C.: Department of Energy. July 17. (DOE, 1996).

Department of Energy. 1997. 1997 D&D Needs. Washington, D.C.: Department of Energy. (DOE, 1997a).

Department of Energy. 1997. Decontamination and Decommissioning Focus Area Quarterly Report, January-March 1997 Activity. Morgantown, WV: Department of Energ, Federal Energy Technology Center. (DOE, 1997c).

Department of Energy. 1997. Decontamination and Decommissioning Technologies List. (http://em52.em.doe.gov/scripts/db...D/Techlist.dbm&first=no&varIndex=0). Washington, D.C.: Department of Energy. (DOE, 1997).

Department of Energy. 1997. Implementation Proposal for Center for Acquisition and Business Excellence. Morgantown, WV: Department of Energy, Federal Energy Technology Center. February 13. (DOE, 1997e).

Department of Energy. 1997. Industry Programs Progress Report. Morgantown, WV: Department of Energy, Federal Energy Technology Center. April. (DOE, 1997).

Department of Energy. 1997. OST Linkage Tables (Predecisional Draft). Washington, D.C.: Department of Energy. October 29. (DOE, 1997).

Department of Energy. 1997. Source Selection Statement for Large-Scale Decontamination and Decommissioning (D&D) Demonstration Projects (Solicitation #2). Washington, D.C.: Department of Energy. April 21. (DOE, 1997).

Department of Energy. 1997. Technology Development Through Industrial Partnerships. Morgantown, WV: Department of Energy, Federal Energy Technology Center. (DOE, 1997).

Department of Energy. 1997. Ten Year Plan Integration (excerpts). Washington, D.C.: Department of Energy. (DOE, 1997).

Department of Energy. undated. Case Histories/Evaluation Committee Review of LSDP Proposals (Solicitation #1). Washington, D.C.: Department of Energy. (DOE, undated).

Department of Energy. undated. Case Histories: Evaluation Committee Review of LSDP Proposals (Solicitation #2). Washington, D.C.: Department of Energy. (DOE, undated).

Department of Energy. undated. D&D Partnership Organizations. Morgantown, WV: Department of Energy, Federal Energy Technology Center. (DOE, undated).

Department of Energy. undated. LSDP Proposals and Their Impact on the 10-Year Plan Activities. Washington, D.C.: Department of Energy. (DOE, undated).

Department of Energy. undated. Technology Screening Examples for CP-5 Concrete Decontamination & Coating Removal. Washington, D.C.: Department of Energy.

Department of Energy. undated. Technology Selection Process and Criteria (Draft) [CP-5]. Washington, D.C.: Department of Energy. (DOE, undated).

Duda, John, Shannon Saget, and Kenneth M. Kasper. 1998. Field Demonstration of Improved D&D Technologies at Hanford's C Reactor. Paper presented at Waste Management '98. (Duda, Saget, & Kasper, 1998).

Hart, Paul. 1997. Decision Making within the Decontamination and Decommissioning Focus Area. Paper presentation to the Committee May 8-9, 1997. (Hart, 1997).

Harness, Jerry. 1997. Presentation to the Committee. Department of Energy. (Harness, 1997).

Hyde, Jerry. 1996. Correspondence From: Jerry Hyde to Carl Bauer and Tom Anderson regarding D&D National Needs. August 1. (Hyde, 1996).

Love, Betty R. 1997. Correspondence to Steven J. Bosart regarding Consensus Reports for the Review Panel of RSI, Institute for Regulatory Science. April 9. (Love, 1997).

Markel, Ken. 1997. Partnering with the Private Sector: The Key to Future Federal R&D Investment. Presentation to Committee on May 8-9. (Markel, 1997).

Strategic Alliance for Environmental Restoration. 1996. Strategic Alliance for Environmental Restoration Overview and Mission. (http://www.strategic-alliance.org). Strategic Alliance for Environmental Restoration. (SAERO, 1996).

Tripp, J.L. 1994. Criteria and Evaluation of Three Decontamination Techniques. Idaho Falls, ID: Department of Energy Operations Office. (Tripp, 1994).

Mixed Waste Focus Area and Other INEEL References

Beitel, Dr. George 1996. A. Mixed Waste Focus Area Program Management Plan. Idaho Falls, ID: Department of Energy, Idaho National Engineering Laboratory. October. (Beitel, 1996).

Beller, John. 1997. Idaho National Engineering Laboratory Environmental Management Technology Development Needs and Opportunities (Draft). LMITCO; Rev. 0.5. Idaho Falls, ID: Department of Energy. (Beller, 1997).

Bonzan, Christine. 1995. Mixed Waste Focus Area Teleconference for FY 96 PEG. June 9. Memorandum for Distribution. (Bonzan, 1995).

Conner, Julie and Mike Connelly. 1996. Mixed Waste Focus Area Briefing to the National Academy of Sciences on November 12. (Conner and Connelly, 1996).

Department of Energy. 1995. Draft Technical Task Plan Proposal Evaluation Criteria--DG-24-95. Idaho Falls, ID: Department of Energy, Idaho National Engineering Laboratory. (DOE, 1995a).

Department of Energy. 1995. EM SSAB Recommendation on the EM FY-98 Integrated Priority List and RDS Rankings. Washington, D.C.: Department of Energy. November 15. (DOE, 1995b).

Department of Energy. 1995. INEL EM Prioritization IPT Report. Idaho Falls, ID: Department of Energy, Idaho National Engineering Laboratory. December. (DOE, 1995d).

Department of Energy. 1995. Scoring Instructions for INEL EM Prioritization. Idaho Falls, ID: Department of Energy/Idaho National Engineering Laboratory. December. (DOE, 1995f).

Department of Energy. 1996. Attachment C--EM SSAB-INEL Meeting Minutes: Recommendation: Fiscal Year 1998 Prioritization of Environmental Management Activities. Washington, D.C.: Department of Energy. January 17. (DOE, 1996a).

Department of Energy. 1996. Charter of the Site Technology Coordination Group (STCG) for the Idaho National Engineering Laboratory. Idaho Falls, ID: Department of Energy. (DOE, 1996b).

Department of Energy. 1996. Environmental Management Fiscal Year 1998 Integrated Budget Prioritization. Washington, D.C.: Department of Energy. March 20. (DOE, 1996e).

Department of Energy. 1996. Final Meeting Minutes Jan. 16-17, 1996, Meeting of the Environmental Management Site Specific Advisory Board--Idaho National Engineering Laboratory Held at the Centre on the Grove in Boise, Idaho: Department of Energy. (DOE, 1996g).

Department of Energy. 1996. INEL Identified Technology Development Needs Prioritization. Washington, D.C.: Department of Energy. (DOE, 1996h).

Department of Energy. 1996. INEL Identified Technology Development Opportunities Prioritization. Washington, D.C.: Department of Energy. (DOE, 1996i).

Department of Energy. 1996. LITCO Director and DOE-ID Program Manager Approval of INEL EM Program, FY97 Murder Board Priority List, ADS Baseline Rev. 4. Idaho Falls, ID: Department of Energy. (DOE, 1996k).

Department of Energy. 1996. Mixed Waste Focus Area Integrated Master Schedule (current as of May 6, 1996). Washington, D.C.: Department of Energy. May.

Department of Energy. 1996. Mixed Waste Focus Area Integrated Technical Baseline Report, Phase 1, Volume 1. Idaho Falls, ID: Department of Energy, Idaho National Engineering Laboratory. January 16.

Department of Energy. 1996. Mixed Waste Focus Area Integrated Technical Baseline Report, Phase 1 Volume 2. Idaho Falls, ID: Department of Energy, Idaho National Engineering Laboratory. January 16.

Department of Energy. 1996. Mixed Waste Focus Area Technology Development Requirements Document, Chemical Oxidation. Washington, D.C.: Department of Energy. July 30. (DOE, 1996l).

Department of Energy. 1996. Mixed Waste Focus Area: Technology Development Requirements Document, Mercury Amalgamation. Washington, D.C.: Department of Energy. July 30. (DOE, 1996n).

Department of Energy. 1996. Mixed Waste Focus Area Technology Development Requirements Document, Mercury Stabilization. Washington, D.C.: Department of Energy. August 19. (DOE, 1996m).

Department of Energy. 1996. Mixed Waste Focus Area Technology Development Transition Guidance. Idaho Falls, ID: Department of Energy, Idaho National Engineering Laboratory.

Department of Energy. 1996. Process for Identifying Technology Needs and Opportunities. Presentation to Committee on November 12-13.

Department of Energy. 1996. Technology and Science Needs Summary (Draft). Washington, D.C.: Department of Energy. (DOE, 1996v).

Department of Energy. 1997. Call for Technology Deployment Initiative (TDI) DOE Proposals. Idaho Falls, ID: Department of Energy, Idaho Operations Office.

Department of Energy. 1997. New DOE-OST Deployment Initiative. Washington, D.C.: Department of Energy. (DOE, 1997g).

Ecology and Environment, Inc. 1996. Overview of Methods Used to Develop the Idaho National Engineering Laboratory: Identified Needs and Opportunities Prioritization Tables (Draft). Idaho Falls, ID: Ecology and Environment, Inc. November 4. (Ecology and Environment, 1996).

Fritz, Lori. undated. Environmental Management Integration at the INEL. Idaho Falls, ID: Department of Energy. (Fritz, undated).

Izatt, R.D. 1996. Memorandum to John Wilcynski regarding Hanford Technology Needs--Mixed Waste Focus Area. January 24. (Izatt, 1996).

Myers, Joy. 1996. Correspondence to John Wilcynski regarding recommendations from the Environmental Management Site Specific Advisory Board. March 25. (Myers, 1996).

Roach, Jay A. 1995. Department of Energy Complex Needs Report: Mixed Waste Focus Area. Idaho Falls, ID: Department of Energy, Idaho Operations Office. November 16. (Roach, 1995).

Snelling, Robert N. and Susan Wood. 1996. Correspondence to Clyde Frank regarding Transmittal of Vitrification Systems Information Package. May 1. (Snelling and Wood, 1996).

Swartz, Ginger. 1995. Taking Stock: An Overview of Public Participation: Lessons Learned by the U.S. Department of Energy, 1990-1995. Boulder, CO: Swartz and Associates. August 1. (Swartz, 1995).

Williams, R. Eric. 1996. Correspondence to Julie Conner regarding Mixed Waste Focus Area High Temperature Melter Strategy Recommendations for FY97. May 31. (Williams, 1996).

Subcon Focus Area and Other Savannah River Site References

Aylward, Robert S., et al. 1996. Technology Needs Statements for Savannah River Site. Aiken, SC: Westinghouse Savannah River Company. (Aylward et al., 1996).

Blundy, Robert P. 1996. Memorandum to Lester Germany regarding Prioritization of ER Needs Statement. May 17.

Brown, James P. 1996. Basis for Investment by the Subsurface Contaminants Focus Area. Paper Presented to Associate Deputy Assistant Secretary for Science and Technology. Aiken, SC: Department of Energy. May 9.

Department of Energy. 1995. Alternative Landfill Cover Backup Data Package for Gate 4 Review. Washington, D.C.: Department of Energy.

Department of Energy. 1995. Management Evaluation Matrix Training Package and Reference Material. Washington, D.C.: Department of Energy. December 4.

Department of Energy. 1996. Alternative Landfill Cover Demonstration: Subsurface Contaminants Focus Area Response to Technical Peer Review Conducted on October 22-23. Albuquerque, NM: Department of Energy.

Department of Energy. 1996. An Independent Panel Evaluation of Organic Treatment and Stabilization Options for Transuranic Wastes at the Savannah River Site. Aiken, SC: Department of Energy. March 26.

Department of Energy. 1996. Documentation and Prioritization of SRS End-User Technology Development Needs. Paper presented to the National Academy of Sciences on December 16. Aiken, SC: Department of Energy, Site Technology Coordination Group.

Department of Energy. 1996. End-User's Guide to Office of Science and Technology Support to the Environmental Restoration Program (Draft). Aiken, SC: Department of Energy, Savannah River Site Technology Coordination Group.

Department of Energy. 1996. National Technology Needs Assessment (Draft). Washington, D.C.: Department of Energy, Landfill Stabilization Focus Area. January 31.

Department of Energy. 1996. National Technology Needs Assessment (Draft). Washington, D.C.: Department of Energy, Containment Plume Containment and Remediation Focus Area. May 17.

Department of Energy. 1996. Original SCFA 13 Prioritization Criteria (Undated Review Draft). Washington, D.C.: Department of Energy. October 3.

Department of Energy. 1996. Priority-Setting/Ranking Criteria. Washington, D.C.: Department of Energy. December 9-10. (DOE, 1996r).

Department of Energy. 1996. Priority-Setting Process for Subsurface Contaminants Focus Area (SCFA) Fiscal Year 1997 Projects. Aiken, SC: Department of Energy, Savannah River Operations Office. June 18.

Department of Energy. 1996. Project Gate Review of Mound Selentec Treatability Study. Washington, D.C.: Department of Energy. April 10.

Department of Energy. 1996. Project Review Process for Contaminant and Remediation (Plumes) Focus Area (Draft). Washington, D.C.: Department of Energy. January 11.

Department of Energy. 1996. Questions/Responses from December 10, 1996 National Academy of Sciences Committee on Environmental Management Technology Meeting with Subsurface Contaminants Focus Area. Aiken, SC: Department of Energy.

Department of Energy. 1996. Report of the Technical Program Review Committee on Non-Destructive Assay (NDA) and Non-Destructive Examination (NDE) Technologies (Predecisional Draft). Washington, D.C.: Department of Energy, Office of Environmental Management. January 25.

Department of Energy. 1996. Savannah River Site Site Technology Coordination Group: General Information Provided to the National Academy of Sciences. (DOE, 1996). Department of Energy. 1996. Priority-Setting/Ranking Criteria. Washington, D.C.: Department of Energy. December 9-10. (DOE, 1996r).

Department of Energy. 1996. Selentec ACT*DE*CON Soil Processing System Data Package, Gate 4 - Entrance to Engineering Development Stage. Washington, D.C.: Department of Energy. October 23.

Department of Energy. 1996. Subsurface Contaminants Focus Area Portfolio Management Decision-Making Process. Washington, D.C.: Department of Energy. .

Department of Energy. 1996. Subsurface Contaminants Focus Area Priority-Setting/Ranking Criteria. Washington, D.C.: Department of Energy.

Department of Energy. 1996. Subsurface Contaminants Focus Area Need Statements and FY 97 Work Packages. Washington, D.C.: Department of Energy. August 22.

Department of Energy. 1996. Subsurface Contaminants Focus Area Program Plan. Washington, D.C.: Department of Energy. October 1.

Department of Energy. 1996. Subsurface Contaminants Focus Area Team Conference Call Minutes. Washington, D.C.: Department of Energy. August 5.

Department of Energy. 1996. Subsurface Contaminants Focus Area, Stakeholder Communication Plan (Predecisional Review). Washington, D.C.: Department of Energy. December 6.

Department of Energy. 1996. Subsurface Contaminants Focus Areas Industry Available Technologies. Washington, D.C.: Department of Energy.

Department of Energy. 1996. Technology Decision Process Pilot Review at Gate 4 of Subsurface Contaminant Focus Area Projects, ACT*DE*CON, Alternative Landfill Cover. Oak Ridge, TN: Department of Energy. October 28-30. (DOE, 1996w).

Department of Energy. 1996. Technology for Wet Storage of Spent Nuclear Fuels. Aiken, SC: Department of Energy. November 12.

Department of Energy. 1996. Technology Screening Concept for the Plumes Focus Area (Summary and refinement of initial report in 1994 by the Plume Focus Area Implementation Team). Washington, D.C.: Department of Energy.

Department of Energy. 1996. Transmittal of D&D Technology Development Needs Statements (U). Aiken, SC: Westinghouse Savannah River Company. December 5.

Department of Energy. 1997. FY96-FY97 Accomplishments. Aiken, SC: Department of Energy, Site Technology Coordination Group.

Department of Energy. 1997. Subsurface Contaminants Focus Area: Response to the National Academy of Sciences-Committee on Environmental Management Technology Visit to the Savannah River Site on December 9-10. Aiken, SC: Department of Energy. January 29.

Department of Energy. undated. Landfills Sites Specification Reports Compilation. Washington, D.C.: Department of Energy.

Department of Energy. undated. Savannah River Site Technology Deployment Initiatives: Building on Success. Aiken, SC: Department of Energy, Site Technology Coordination Group.

Department of Energy. undated. Technology Development Program. Aiken, SC: Department of Energy, Savannah River Site Technology Coordination Group.

Hicks, Tom et al. 1996. Work Package Prioritization for the Subsurface Contaminants Focus Area. Prepared by a panel representing the United States Department of Energy Savannah River including the Plumes Focus Area and the Landfill Stabilization Focus Area, the Waste Policy Institute and the Savannah River Technology Center. Aiken, SC: Department of Energy. (Hicks et. al., 1996).

Hudson, P.I. et al. 1996. Transmittal of Solid Waste Division Technology Development Needs. Aiken, SC: Westinghouse Savannah River Company. December 5. (Hudson, 1996).

Wright, Jim. 1996. Priority-Setting/Ranking Criteria Presentation to the Committee. National Academy of Sciences, Committee on Environmental Management Technologies, Decision-Making Process for Selection and Prioritization Panel. (Wright, 1996).

Tanks Focus Area and Other Hanford Site References

Department of Energy. 1992. Hanford Strategic Plan. Richland, WA: Department of Energy. (DOE, 1992).

Department of Energy. 1996. Tanks Focus Area Multiyear Program Plan FY97-FY99. Richland, WA: Department of Energy/Pacific Northwest Laboratory. August. (DOE, 1996u).

Department of Energy. 1996. Technology Needs/Opportunities Statement Outline (Hanford Site). Richland, WA: Department of Energy. December 13. (DOE, 1996y).

Department of Energy. 1997. Technology Needs Statements. Washington, D.C.: Department of Energy. (DOE, 1997l).

Department of Energy. undated. Management Plan. Richland, WA: Department of Energy. (DOE, undated-a).

Department of Energy. undated. Mixed Waste Subgroup Technology Needs Scoring Sheet. Richland, WA: Department of Energy. (DOE, undated-b).

Department of Energy. undated. Problem Prioritization Criteria (Sample). Richland, WA: Department of Energy.

Department of Energy. undated. Tank Technology Needs Scoring Sheet (Sample). Richland, WA: Department of Energy.

Department of Energy, Environmental Protection Agency, Washington State Department of Ecology, Pacific Northwest National Laboratory, Tri-Cities Commercialization Partnership. 1996. Hanford Technology Deployment Center Program Plan. Richland WA: Department of Energy. August. (DOE et al., 1996).

Frey, Jeff and Tom Brouns. 1997. Tanks Focus Area Process for Program Development Needs Identification & Technical Response Development. Richland, WA: Department of Energy/Tanks Focus Area. January 9. (Frey & Brouns, 1997).

Saget, Shannon. 1996. Correspondence from Shannon Saget to Joe Paladino regarding Hanford STCG Prioritization Criteria and Site Needs. February 29. (Saget, 1996).

Saget, Shannon. 1996. Summary--Hanford STCG Prioritization Methodology. Richland, WA: Department of Energy. (Saget, 1996).